5th International Conference on Smart Material Research (ICSMR 2019)

IOP Conference Series: Materials Science and Engineering
Volume 856

Brisbane, Australia
13 - 16 November 2019

ISBN: 978-1-7138-2686-6
ISSN: 1757-8981

Printed from e-media with permission by:

Curran Associates, Inc.
57 Morehouse Lane
Red Hook, NY 12571

Printed with permission by Curran Associates, Inc. (2021)

For permission requests, please contact the Institute of Physics
at the address below.

Institute of Physics
Dirac House, Temple Back
Bristol BS1 6BE UK

Phone: 44 1 17 929 7481
Fax: 44 1 17 920 0979

techtracking@iop.org

Additional copies of this publication are available from:

Curran Associates, Inc.
57 Morehouse Lane
Red Hook, NY 12571 USA
Phone: 845-758-0400
Fax: 845-758-2633
Email: curran@proceedings.com
Web: www.proceedings.com

TABLE OF CONTENTS

MATERIAL MECHANICS

COMPOSITE MATERIALS

IOP Conf. Series: Materials Science and Engineering **856** (2020) 011001 doi:10.1088/1757-899X/856/1/011001

PREFACE

It is with great pleasure that I introduce the proceedings of the 2019 5th International Conference on Smart Material Research (ICSMR 2019) which was held recently in Brisbane, Australia, 13-16, November 2019. ICSMR 2019 was dedicated to technical issues related to smart material research.

The objectives of the conference were not only to provide opportunities to the international scientists to present state of the art research in smart materials but also to provide a forum to discuss ideas for future directions in the field and explore possibilities for collaborative research projects.

The conference program included keynote, oral, and poster presentations from scholars working in the areas of materials science and engineering from all over the world. It covered recent trends and progress made in the field of materials engineering and manufacturing. Professors from Australia and USA were invited to deliver four keynote lectures regarding the latest information in their respective areas of expertise.

The proceedings present a selection of high quality papers submitted to the conference by researchers from universities, research institutes, and industry. All papers were subjected to peer-review by conference committee members and international reviewers. The papers were selected based on their quality and their relevance to the conference. The proceedings present recent advances in the fields of materials science and engineering and manufacturing as well as in other related areas including materials properties, measuring methods, analytical and numerical modeling and simulations among others.

I would like to express special gratitude to members of the conference committee and organizers of the conference. I would also like to thank the reviewers for their valuable time and advice which helped in improving the quality of the papers selected for presentation at the conference and for publication in the proceedings. Finally I want to thank the authors, the members of the organizing committee, the reviewers, the chairpersons, sponsors, and all other conference participants for their support of ICSMR 2019.

Prof. Ramesh K. Agarwal
Washington University in St. Louis, USA
January 21, 2020

IOP Conf. Series: Materials Science and Engineering **856** (2020) 011002 doi:10.1088/1757-899X/856/1/011002

Conference Committee Chairs

Darren Martin, University of Queensland, Australia
Ramesh K. Agarwal, Washington University in St. Louis, USA

Program Committee Chairs

Jiazhao Wang, University of Wollongong, Australia

International Technical Committees

Agus Arsad, Universiti Teknologi Malaysia, Malaysia
Ai-Bao Chai, University of Nottingham Malaysia Campus, Malaysia
Alaa El-Din Ali El-Shafei, Mansoura University, Mansoura, Egypt
Blanka Tomková, Technical University of Liberec, Czech Republic
Carlos Rolando Rios-Soberanis, Centro de Investigación Científica de Yucatán, México
Chi-wai Kan,The Hong Kong Polytechnic University, Hong Kong
Dauda Shehu Ibrahim, Bayero University, Nigeria; Southeast University, China
Dongbin Wei, University of Technology Sydney (UTS), Australia
G. RAJARAJAN, Hindustan Institute of Technology & Science (Deemed to be University), India
Guang-Min Luo, National Kaohsiung University of Science & Technology, Taiwan
Hideaki Tsukamoto, Hosei University, Japan
Irshad Ahmad, American University of Ras Al Khaimah, UAE
Juliana Anggono, Petra Christian University, Indonesia
Lilly Mercy, Sathyabama University, India
Mochamad Chalid, Universitas Indonesia, Indonesia
Muhammad Abbas Ahmad Zaini, Universiti Teknologi Malaysia, Malaysia
Nikoloz Jalabadze, Georgian Technical University in Tbilisi, Georgia
Nilofar Asim, National University of Malaysia, Malaysia
Norhazilan Md. Noor, Universiti Teknologi Malaysia, Malaysia
P Bílý, Czech Technical University in Prague, Czechia
Patricia Sanmartín Sánchez, Universidade de Santiago de Compostela, Spain
Patrice Berthod, University of Lorraine, France
Redzuan Abdullah, Universiti Teknologi Malaysia, Malaysia
Siti Rahmah Aid, Malaysia-Japan International Institute of Technology(MJIIT),Universiti Teknologi Malaysia, Malaysia
Theodoros C. Rousakis, Democritus University of Thrace (D.U.Th.), Greece
Vijay Sekar K.S., SSN College of Engineering, India
Xiangjiang Liu, Zhejiang University, China
Yanisa Laoong-u-thai, Burapha University, Thailand
Yu-Cheng Chen, EPS bio technology corp., Taiwan
Zahira Yaakob, Universiti Kebangsaan Malaysia, Malaysia

IOP Conf. Series: Materials Science and Engineering **856** (2020) 011003 doi:10.1088/1757-899X/856/1/011003

First name	Surname	Institution	E-mail address
Patrice	Berthod	University of Lorraine, France Institut	patrice.berthod@univ-
Charles Lois Ildesa	Flores	University of the Philippines, Philippines	ciflores@up.edu.ph
Neda Ghaebi	Panah	The University of Western Australia,	Neda.ghaebipanah@g
Dauda Shehu	Ibrahim	Southeast University, China	233179947@seu.edu.
Lusia Silfia Pulo	Boli	Institut Teknologi Bandung, Indonesia;	lusiasilfia@gmail.com
Vera	Khoirunisa	Institut Teknologi Bandung, Indonesia;	vera.khoirunisa@tf.ite
Hideaki	Tsukamoto	Hosei University, Japan	htsukamoto@hosei.ac.
Nadia	Nassif	University of Sharjah, United Arab	u17105811@sharjah.a
Anita Setyowati Srie	Gunarti	Universitas Islam 45, Indonesia	anitassgunarti@gmail.
Ratnaningsih Eko	Sardjono	Universitas Pendidikan Indonesia,	ratnaeko@upi.edu;

IOP Conf. Series: Materials Science and Engineering **856** (2020) 011004 doi:10.1088/1757-899X/856/1/011004

Peer review statement

All papers published in this volume of *IOP Conference Series: Materials Science and Engineering* have been peer reviewed through processes administered by the proceedings Editors. Reviews were conducted by expert referees to the professional and scientific standards expected of a proceedings journal published by IOP Publishing.

IOP Conf. Series: Materials Science and Engineering **856** (2020) 012001 doi:10.1088/1757-899X/856/1/012001

Punching Shear Strength of Reinforced Concrete Flat Slabs with Macro Synthetic Fibers

Nadia Nassif [1], Salah Altoubat [1], Mohamed Maalej [1] and Pierre Estephane [2]

[1] Civil and Environmental Engineering Department, University of Sharjah, Sharjah, United Arab Emirates – 27272.
[2] GCP Applied Technologies, 5006, Dubai, United Arab Emirates.

u17105811@sharjah.ac.ae

Abstract. Discrete macro fibers such as steel and synthetic have been recognized as an attractive material to improve concrete properties such as flexural toughness, shear strength, impact characteristics, deflection capacity and concrete ductility. In this paper, the effect of macro synthetic fibers on punching shear resistance and cracking behaviour of reinforced concrete slabs was investigated. A total of six reinforced concrete slabs were designed, instrumented and tested in displacement control mode, under monotonic center-point load in simply supported configuration. Synthetic fibers were added to the concrete slabs at dosage rates of 0.5% and 1% by volume. The results have shown a significant increase of the punching shear strength, energy absorption capacity and ductility of the slabs. The addition of 0.5% and 1% of macro synthetic fiber to the slab increased punching shear strength by 30% and 70% and increased energy absorbance by 40% and 80%, respectively.

1. Introduction

Punching shear failure is a type of failure in reinforced concrete slabs subjected to high localized forces. This type of failure is generally very brittle and occurs suddenly without warning. Punching failure can be prevented using traditional reinforcement methods such as shear stirrups, studs or drop panels. Integrating discrete macro fibers into concrete mix has been recognized as an effective method to improve the punching strength of concrete slabs [1-4]. The use of synthetic fibers increase the punching capacity, improve the energy absorption performance of the column–slab connection and transform brittle-type punching shear failure into a gradual and ductile shear failure [2,3]. These fibers are manufactured from polymer-based materials such as nylon, polyethylene and polypropylene. Synthetic fibers are added during or before mixing process. Synthetic fibers are highly durable, have lesser unit weight and unlike steel fibers, they are non-corrosive. Previous experiments in literature indicate that fibers contribution to punching shear strength lies in their bridging effect to limit crack propagation and minimize their width at critical sections. This bridging effect considerably enhances the post-cracking tensile strength at the tensile zone. It is believed that this improvement is basically attributed to synthetic fibers tensile and bonding strength with concrete. In recent years, many research papers are published focusing on punching shear strength in flat slabs containing steel fibers [5,6].
Utilization of steel fibers to enhance punching shear strength of flat plates has been introduced by Swamy and Ali, (1982) [5].They reported enhancement in maximum shear load capacity between 23%

to 42% corresponding to adding steel fiber volume ratio contents from 0.6 percent (48 kg/m^3) to 1.2 percent (96 kg/m^3). In 1995, M. Harajli [6] tested twelve small-scaled slab-column specimens. His study results showed that adding steel fibers up to 2.0 percent (160 kg/m^3) of the total volume ratio increased the ultimate punching capacity by about 36 percent compared with plain concrete slabs. However, changing thickness had no noticeable effect on the ultimate punching capacity. In 2010 Cheng and Gustavo [7] tested 10 slabs. Their results have shown that the addition of steel fibers led to an increase in punching shear capacity and deflection capacity particularly when 1.5 % of fibers were used. Moreover, the results showed that slabs with lower flexural reinforcement ratio exhibited a change in failure mode from punching shear to flexural yielding. In 2012 L. Nguyen-Minh, M. Rovňák, and Toan Tran-Quoc [8] tested 12 flat slabs. The fiber content was (30, 45, 60 kg/m^3). They concluded that using steel fibers increases punching shear capacity up to 39%. It also reduces the average crack width up to 40% considering serviceability limit state. Furthermore, up to 36% reduction of deflection was observed. It was noticed that fibers shape and length to diameter ratio should be taken into account in calculations of the punching shear capacity. In 2016 Ahmed Abdel-Rahman, Nasr Hassan and Adel Soliman [9] tested 14 slab-column connections. Their results showed that using steel fiber increased both the failure load and energy absorbing capacity. It was found that the higher the steel fiber ratio was the higher the ultimate load carrying capacity of the specimens. Increasing the steel fiber by 1.5% enhanced the ultimate capacity of the specimens by 24%. In 2018 T.musse and E. liberati [10] tested 8 square slabs. The fiber percentage used was 0.9%, which is equivalent to 70.6 kg of fibers per cubic meter of concrete. Their results have shown that the use of fibers increased ultimate loads by 26% for the slab without shear reinforcement and increased approximately 12% for the slabs with shear reinforcement.
Research on punching shear strength of slabs with synthetic fibers however, is still scarce in the existing literature and thus more experimental work is needed which motivates the current study. The current paper presents and discusses experimental results obtained by testing to failure simply supported slabs containing macro synthetic fibers under center point loading.

2. Experimental program

2.1. *Materials*

The concrete mixture proportions and properties used in casting the six slabs are provided in Table 1. The concrete mix was designed to achieve a 28-day compressive strength of 30 MPa. Ordinary Portland cement was used with dune sand and aggregates having (5 mm, 10 mm, 20 mm) maximum size. CONCERA™ SA8476 was used as an admixture, it is a high efficiency polycarboxylate-based superplasticizer intended for the production of Control Flow Concrete, a highly flowable concrete utilizing conventional mix designs with extended slump-flow life and high resistance to segregation. Test slabs were reinforced with 12 mm diameter rebar of 420 MPa yield strength.

Table 1. Mix design used in the current study

Unit Weight (Kg/m^3)									
Mix	OPC	Aggregates			Dune Sand	Free W/C	Total water	Fibers	CONCERA
		20mm	10mm	5mm					
Control	360	601	271	581	372	0.58	227	0	0.5
0.5% Fiber	360	601	271	581	372	0.58	227	4.6	3
1% Fiber	360	601	271	581	372	0.58	227	9.2	3.67

The macro synthetic fiber used in this testing program is "STRUX ® 90/40" (Figure 1) which is a commercial product used in many industrial applications .The main components of this polymeric fiber type are polypropylene and polyethylene The length of individual fiber is 40 mm with aspect ratio of 90. "STRUX ® 90/40" role in concrete mix is purely mechanical and will not interfere with cement hydration. Table 2 shows "STRUX ® 90/40" properties.

Table 2. "STRUX ® 90/40" Properties

Parameter	Value
Specific gravity	0.92
Absorption	None
Modulus of elasticity	1,378 ksi (9.5 GPa)
Tensile strength	90 ksi (620 MPa)
Alkali, acid & salt resistance	High

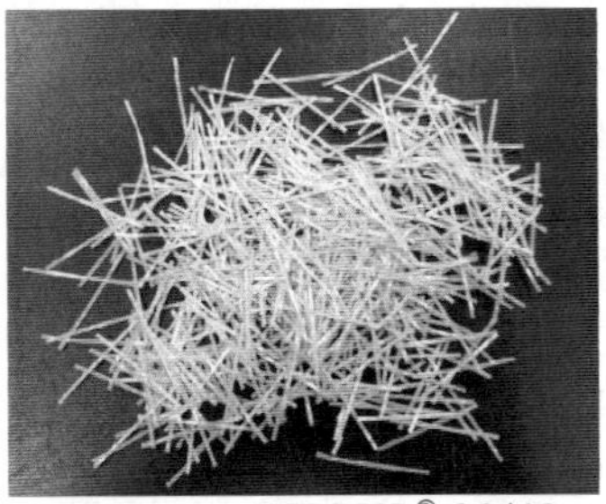

Figure 1. STRUX ® 90/40

2.2. Slab specimens and testing procedure

A total of 6 slabs were tested. All slabs are of the same thickness of 100 mm and tensile reinforcement ratio $\rho = 0.9$ %. The slab dimensions were the same for all six specimens, 1.5 m x 1.5 m with a 100 mm square column stub at the centre of the slab for load application. The slabs were divided into three groups with respect to their fiber content as follows: 2 plain concrete slabs with no fiber; 2 slabs with 0.5 % of macro synthetic fiber and 2 slabs with 1 % of macro synthetic fiber. The addition of fibers by up to 1% is commonly and practically used in industrial applications and previous studies. Duplicate slabs were tested to obtain average response and to assess the reliability and reproducibility of the test. Slabs were cast and cured under similar conditions and tested after 28 days. Measured slumps of synthetic fiber reinforced concrete and plain concrete were 160 mm and 145 mm, respectively. Table 3 provides details of slabs tested in this study.

Table 3. Slabs details

Slabs	Quantity	Dimension (mm)	Thickness (mm)	Effective depth (mm)	Steel ratio %	Fiber %	Compressive strength (Mpa)
Group 1	2	150 x 150	100	75	0.9	0	28.0
Group 2	2	150 x 150	100	75	0.9	0.5	29.0
Group 3	2	150 x 150	100	75	0.9	1	29.45

All slabs were tested under center point load. The load was applied monotonically in displacement control mode at a rate of 0.015 mm/sec. A vertically oriented hydraulic actuator connected to a steel reaction frame was used for application of the load to the slab specimens, as shown in Figure 2. All slabs tested were simply supported along four edges. Linear Variable Displacement Transducers (LVDTs) were placed under the slab in mid-span to monitor the deflection. At each load level, vertical deflection and crack patterns on the bottom surface of all slabs were recorded.

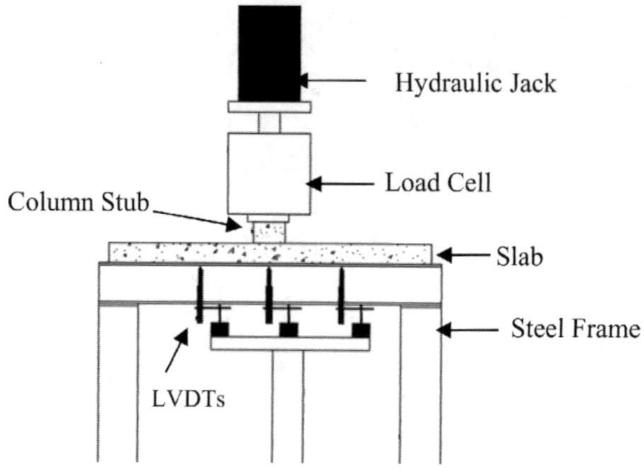

Figure 2. Test setup

3. Results and discussions

3.1. Cracks and failure mode

Punching shear failure was primarily occurred in all the tested slabs. This was clearly evidenced by the column stub punch through the slab top surface and defined by the cracking patterns in the bottom of the slab surface. Visual observation of cracks intensity provides qualitative assessment of slabs punching behaviour. Slabs without fibers failed abruptly and with little cracking intensity in the punching area, and some parts of the bottom concrete cover fell apart (Figure 3a).

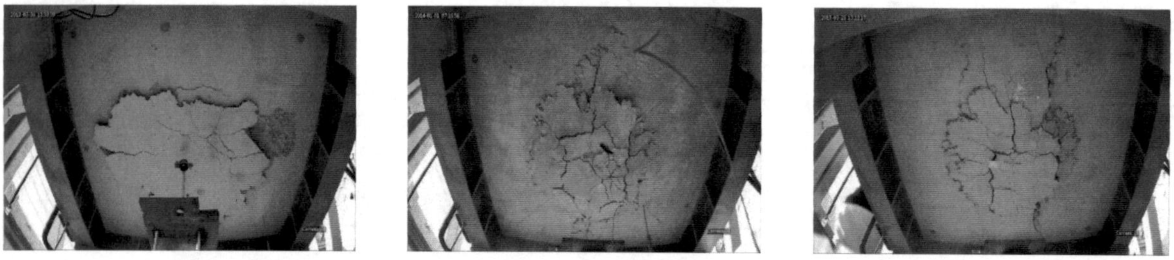

(a) Slab with 0% Fibers **(b)** Slab with 0.5% Fibers **(c)** Slab with 1.0% Fibers

Figure 3. Observed cracking patterns on the bottom surface of slabs

Unlike the slabs without fibers, slabs with fibers have failed in more ductile manner, the punching area became more defined and smaller than slabs with no fibers due to bridging effect of fibers (Figures 3b and 3c). Moreover, the crack intensity increased as the percentage of fibers increased. Furthermore, the slabs with 1% fiber have shown signs of flexural cracks which indicates that fibers increased punching shear capacity to a level where flexural yielding is induced (Figure 3c).

3.2. Peak Loads

Punching load and slab deflection were recorded during the test. Load-deflection curves for all slabs are presented in Figure 4. The results in Figure 4 show that fibers increased the punching load capacity of the slab. Adding 0.5% and 1% of macro synthetic fibers to the slab increased the peak punching load (P_u) by 30% and by 70% respectively relative to control slabs as shown in Figure 5. The average peak load of the duplicate slabs is reported in Figure 5 and summarized in Table 4.

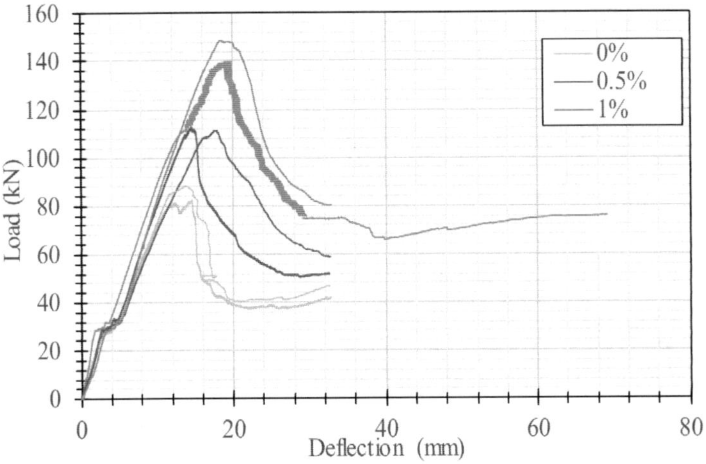

Figure 4. Load versus deflection response for all tests

Table 4. Summary of test results

Fiber	Average Peak Load (kN)	Average Energy absorbed (kJ)
0%	85.22	1898.89
0.5%	112.15	2518.74
1%	143.93	3400.97

3.3. Energy Absorbed

Energy absorption is defined by the total area under the load-deflection curve. The areas under the curves were computed. To preform consistent comparison, the area under the load deflection curves was calculated up to 40 mm central deflection at which the control slabs completely failed. The ductility of control slabs was much lower than the slabs with fibers which failed at 75mm deflection. The computed absorbed energy (E) was normalized relative to the control slab (E_c) and illustrated in Figure 6. The result shows that macro synthetic fibers at 0.5% and 1% fibers increased the absorbed energy (relative to control) by 40% and 80%, respectively.

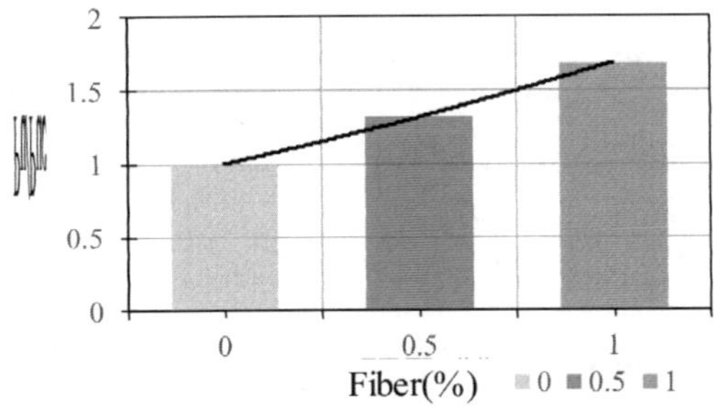

Figure 5. Relative average punching force capacity

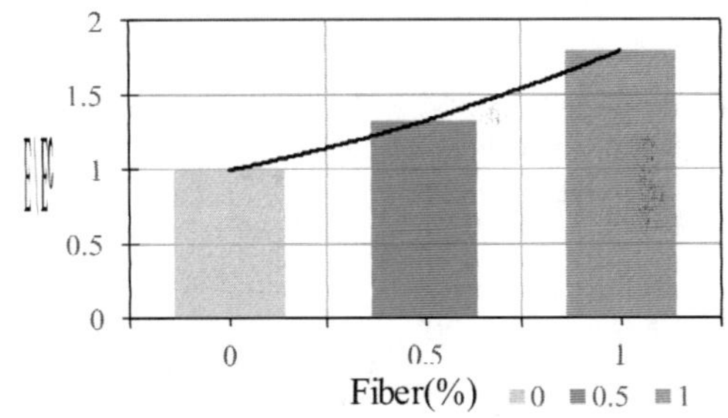

Figure 6. Relative average energy absorption capacity

3.4. Comparison with previous studies

The results obtained from 62 full scale tests of slabs with steel fibers reported in the literature by Higashiyama [4] was compiled and used in the analysis. Minitab®18 software was used to establish a linear regression model for the compiled results. The following regression model (equation 1) with a mean square error (R^2) of 0.89 was obtained as follows:

$$P_u = -361.2 - 0.71t + 4.13d + 46.2\rho_s + 58.6\rho_f + 1.341L + 1.155fc' \quad (1)$$

Where: P_u is the maximum punching load for the slab in kN. t and d are the slab thickness and effective depth, respectively in mm. ρ_s and ρ_f are the steel and fiber percentages respectively. L is the loading plate length in mm, and fc' is the compressive strength of concrete. The model developed for steel fibers in equation (1) was used to predict the punching shear capacity of slabs with macro synthetic fiber. Experimental and predicted values by the regression model are displayed in Figure 7.

The results shows that the model for steel fibers predict the results with synthetic fibers reasonably well. Furthermore, the current test results fit within the previous studies done on steel fibers.

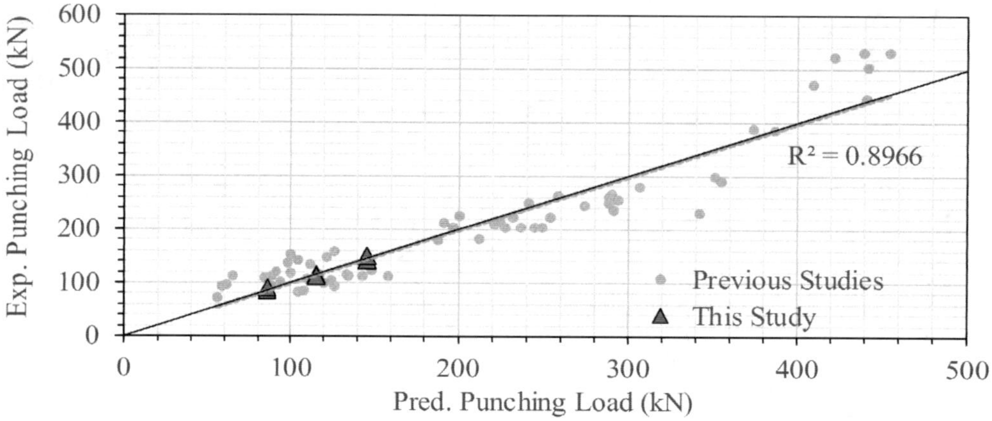

Figure 7. Experimental and predicted results for current and previous studies.

4. Conclusion

This paper discussed the enhancement of punching shear strength in concrete slabs using macro synthetic fibers. A total of six slabs with thickness 100 mm were tested: 2 plain concrete slabs with no fiber, 2 slabs with 0.5% fibers and 2 slabs with 1% fibers. Slabs were tested under center point load in displacement control mode. The results have revealed a significant improvement in the punching shear capacity, energy absorption and ductility of the reinforced concrete slab. The addition of 0.5% and 1% of macro synthetic fiber to the slab increased punching shear strength by 30% and 70% and increased energy absorbance by 40% and 80%, respectively. Slabs without fibers behaved in a brittle way whereas slabs with fibers have failed in more ductile manner. The load-deflection curves of the tested slabs showed significant increase in ductility due to the addition of macro synthetic fibers. However, different slab geometries and boundary conditions shall be further considered in future studies to better assess and evaluate punching shear behavior of fiber reinforced concrete slabs.

References

[1] Nguyen-Minh L, Rovnák M, Tran-Quoc T, Nguyenkim K. Punching shear resistance of steel fiber reinforced concrete flat slabs. *Procedia Engineering*. 2011 Jan 1;14:1830-7.

[2] Roesler JR, Altoubat SA, Lange DA, Rieder KA, Ulreich GR. Effect of synthetic fibers on structural behavior of concrete slabs-on-ground. *ACI materials journal*. 2006;103(1):3.

[3] Roesler, J. R., Lange, D. A., Altoubat, S. A., Rieder, K. A., & Ulreich, G. R., 2004. Fracture of plain and fiber-reinforced concrete slabs under monotonic loading. Journal of Materials in Civil Engineering, 16(5), 452-460.

[4] Higashiyama H, Ota A, Mizukoshi M. Design equation for punching shear capacity of SFRC slabs. *International Journal of Concrete Structures and Materials*. 2011 Jun;5(1):35-42.

[5] Swamy RN, Ali SA. Punching shear behavior of reinforced slab-column connections made with steel fiber concrete. In *Journal Proceedings* 1982 Sep 1, Vol. 79, No. 5, pp. 392-406.

[6] Harajli MH, Maalouf D, Khatib H. Effect of fibers on the punching shear strength of slab-column connections. *Cement and Concrete Composites*. 1995 Jan 1;17(2):161-70.

[7] Cheng MY, Parra-Montesinos GJ. Evaluation of Steel Fiber Reinforcement for Punching Shear Resistance in Slab-Column Connections--Part I: Monotonically Increased Load. *ACI Structural Journal*. 2010 Jan 1;107(1).

[8] Nguyen-Minh L, Rovňák M, Tran-Quoc T. Punching Shear Capacity of Interior SFRC Slab-Column Connections. Journal of Structural Engineering. 2011 Aug 22;138(5):613-24.

[9] Abdel-Rahman AM, Hassan NZ, Soliman AM. Punching shear behavior of reinforced concrete slabs using steel fibers in the mix. HBRC journal. 2018 Dec 1;14(3):272-81.
[10] Musse TH, LIBERATI E, TRAUTWEIN L, GOMES R, Guimarães GN. Punching shear in concrete reinforced flat slabs with steel fibers and shear reinforcement. Revista IBRACON de Estruturas e Materiais. 2018 Oct;11(5):1110-21.

IOP Conf. Series: Materials Science and Engineering **856** (2020) 012002 doi:10.1088/1757-899X/856/1/012002

Mechanical Properties Improvement of Clays Using Silica Sand Waste and Dust Sand Foundry Waste

Anita Setyowati Srie Gunarti[1,*], Irwan Raharja[2]

[1] Department of Civil Engineering, Universitas Islam 45, Jl. Cut Meutia No. 83, Bekasi, West Java, Indonesia
[2] Department of Accounting Information Systems, Universitas Bina Sarana Informatika, Jl. Banten No. 1, Karangpawitan, Karawang Barat, West Java, Indonesia

anita_ft@unismabekasi.ac.id

Abstract. The increase of industrial waste and many road damages due to unstable subgrade invite the research on soil additives made from waste treatment as stabilizers to improve the mechanical value of clay. This study used California Bearing Ratio Test based on the ASTM standard for mechanical properties testing of clay. The soil stabilization method was carried out by mixing the clay with Silica Sand waste and Dust Sand Foundry waste in a predetermined composition. California Bearing Ratio value of 2.5 mm penetration showed a significant increase of 47.9% from the original clay in the composition of the waste 5% Silica Sand and 5% Dust Sand Foundry.

1. Introduction

A mineral factory in the Jababeka Industrial Estate, Cikarang, West Java, produces 50 tons of silica sand (SS) waste every month. Another factory, a metal mold manufacturing industry, produces the Dust Sand Foundry (DSF) waste. Both of these wastes contain compound elements that can be used as stabilizers or clay soil additives. These new materials are cheap and available for improving the subgrade. Some researchers have conducted research using waste or other additives, yet research by mixing two types of SS and DSF waste is still lacking. The CBR value showed an upward trend in stabilizing the soils using Ground Granulated Blast Furnace Slag (GGBS) industrial waste [1]. Another study utilized the recycled tiles and tires in stabilizing the soils which gave a good effect as a substitute for cement or sand [2]. The use of electric arc furnace dust (EAFD), an industrial by-product in lieu of cement or lime, showed a significant improvement in mechanical properties and durability of sand [3]. The stabilizing effect of powdered glass on clay soil showed an increase in CBR values in the soil with the addition of 5% glass powdered [4]. Addition of Saw Dust Ash to clayey soil, resulting in increased CBR and unconfined compressive strength [5]. Garnet waste as much as 40% mixed on the ground for road shoulder showed an improvement in the CBR value [6].

This research proposed an innovative combination of two different types of industrial waste, namely SS and DSF, as additives to replace cement or lime for clay soil stabilization. Improving the mechanical properties based on California Bearing Ratio (CBR) values in soil mixtures using SS and DSF wastes, and finding the optimal waste composition with maximum CBR values were the aim of this study. This study did not review the changes in soil chemical structure due to soil stabilization using waste as the limitation but only focused on mechanical changes in soil stabilized through CBR values. The research contributes to the report of a method of combining two different types of industrial wastes, namely SS and DSF as clay soil stabilization additives in order to increase the CBR value and to find the composition/percentage of waste based on the highest increase in CBR value. Potential Impact of this research is the discovery of compositions and combinations of materials made from industrial waste that are useful as additives for improving the carrying capacity of clay in subgrade roads. In addition, this waste utilization will produce technical and economic benefits.

2. Materials and Methods

The material in this research was the clay gathered from the soil of Universitas Islam 45 Bekasi at a depth of 20-50 cm from the original ground surface (disturbed condition), Silica Sand (SS) waste, and Dust Sand Foundry (DSF) waste. Samples were prepared by mixing air-dried clay conditions with SS waste and DSF waste as additives or stabilizers, then left in a closed bag for seven days at room temperature. The composition of the waste was as follows: SS waste 2.5% and 5% of the soil dry weight, and DSF waste as much as 2.5% and 5% of the soil dry weight. The mechanical properties test of the un-soaked clay is California Bearing Ratio (CBR) with 2.5 mm penetration following the ASTM standard. This test was carried out under Maximum Dry Density (MDD) and Optimum Moisture Content (OMC) conditions based on the results of the proctor standard compaction test. Variations in waste composition are shown in table 1.

Table 1. Additives Composition

Varian Number	Additives	
	SS[a] (%)	DSF[b] (%)
1[c]	0	0
2	0	2.5
3	0	5
4	2.5	2.5
5	2.5	5
6	5	2.5
7	5	5

[a]Silica Sand Waste
[b]Dust Sand Foundry Waste
[c]Original Clay

3. Result and Discussion

Clay gathered from Universitas Islam 45 Bekasi has a specific gravity value (Gs) of 2.510 [7]-[8] and is categorized as inorganic type clay [9]. CBR test results for all variants are shown in table 2. CBR values for all variants of additives composition showed an increase in the CBR value of the original clay, whereas the OMC value showed a decrease in the OMC value on the original clay. This is due to the very fine-grained waste filling the pore cavity of the soil, which in the original clay, the cavity is filled with water and air [10]. The water content in the soil decreases due to the presence of fine grains of waste that fill the pore cavity. An increase in the number of solid particles in the soil results in an increase in the weight of the dry volume compared to the original clay soil conditions.

Table 2. Mechanical Properties Value

Variant Number	Additives Composition		OMC [c] (%)	CBR [d] (%)
	SS[a] (%)	DSF[b] (%)		
1	0	0	30.50	7.14
2	0	2.5	29.15	9.24
3	0	5	29.10	9.68
4	2.5	2.5	30.135	7.91
5	2.5	5	29.04	10.11
6	5	2.5	29.78	9.01
7	5	5	29.08	10.56

[a]Silica Sand Waste
[b]Dust Sand Foundry Waste
[c]Optimum Moisture Content
[d]California Bearing Ratio

Figure 1 shows the increase of the CBR value based on the increase the DSF waste percentage. Chemical reactions occur in the mixture of soil and waste during curing 7 days, thus forming new grains that were harder and stronger to withstand the load. Waste mixed with clay caused the cation exchange process from the soil to be replaced by cation from the waste so that the size of the clay granules increases (flocculation). In addition to the flocculation process that occurs in soil stabilization, there was also a hydration process.

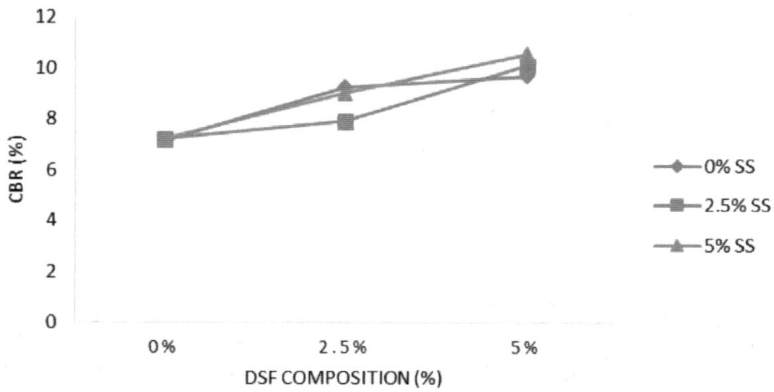

Figure 1. CBR Value For The addition of DSF waste

Figure 2 shows that the CBR value tends to increase following the percentage of SS waste. It can be seen at the 0% DSF line, and the 5% DSF. In the other hand, the 2.5% DSF line shows the decrease of the CBR value when the percentage of SS waste is decreased.

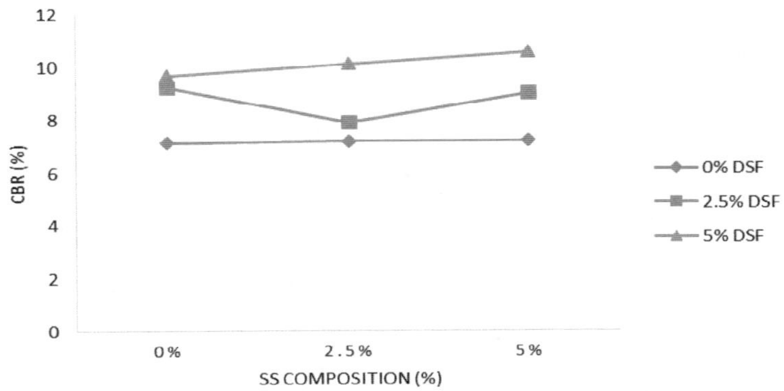

Figure 2. CBR Value For The addition of SS waste

4. Conclusions

Based on the CBR value, clay mixed with SS waste and DSF waste showed an increase in mechanical properties. The highest CBR value was shown in the clay mixture with an additives composition of 5% SS waste and 5% DSF that showed the increased of CBR value of 47.9% compared to of the original clay. Further research is needed on testing the chemical properties of stabilized clay soils using SS and DSF, as well as environmental feasibility testing.

Acknowledgements

The authors thank to DRPM RISTEKDIKTI for funding this research in the 2019 Applied Research Grant Program based on SK No. T/140/E3/RA.00/2019 and SK No. 226/SP2H/LT/DRPM/2019.

References

[1] Manimaran, A., Santhosh, S. and Ravichandran, P.T., 2018. Characteristics Study on Sub Grade Soil Blended with Ground Granulated Blast Furnace Slag. *Rasayan Journal of Chemistry*, *11*(1), pp.401-404.

[2] Al-Bared, M.A.M., Marto, A. and Latifi, N., 2018. Utilization of recycled tiles and tyres in stabilization of soils and production of construction materials–A state-of-the-art review. *KSCE Journal of Civil Engineering*, *22*(10), pp.3860-3874.

[3] Al-Amoudi, O.S.B., Al-Homidy, A.A., Maslehuddin, M. and Saleh, T.A., 2017. Method and mechanisms of soil stabilization using electric arc furnace dust. *Scientific reports*, *7*, p.46676.

[4] Olufowobi, J., Ogundoju, A., Michael, B. and Aderinlewo, O., 2014. Clay soil stabilisation using powdered glass. *Journal of Engineering Science and Technology*, *9*(5), pp.541-558.

[5] Butt, W.A., Gupta, K. and Jha, J.N., 2016. Strength behavior of clayey soil stabilized with saw dust ash. *International Journal of Geo-Engineering*, *7*(1), p.18.

[6] Sani, W.M., Mohamed, A., Nor, H.M., Kamarudin, N.A.S., Khalid, N.H.A., Sam, A.M. and Ab Rashid, R., 2019, January. Characterization of soil mixed with garnet waste for road shoulder. In *IOP Conference Series: Earth and Environmental Science* (Vol. 220, No. 1, p. 012052). IOP Publishing.

[7] Kholis, N., Gunarti, A.S.S. and Sylviana, R., 2018. Stabilisasi Tanah Lempung Menggunakan Semen dan Renolith. *BENTANG: Jurnal Teoritis dan Terapan Bidang Rekayasa Sipil*, *6*(1), pp.62-77.

[8] Gunarti, A.S.S., 2014. Daya Dukung Tanah Lempung yang Distabilisasi dengan Spent Catalyst RCC 15 dan Kapur. *BENTANG: Jurnal Teoritis dan Terapan Bidang Rekayasa Sipil*, *2*(1), pp.38-52.

[9] Hardiyatmo, H.C., 2012. Mekanika Tanah 1, Edisi

[10] Andriani, A., Yuliet, R. and Fernandez, F.L., 2012. Pengaruh Penggunaan Semen sebagai Bahan Stabilisasi pada Tanah Lempung Daerah Lambung Bukit terhadap Nilai CBR Tanah. *Jurnal Rekayasa Sipil (JRS-Unand)*, *8*(1), pp.29-44.

IOP Conf. Series: Materials Science and Engineering **856** (2020) 012003 doi:10.1088/1757-899X/856/1/012003

The effect of drying method on the surface structure of mesoporous sol-gel derived bioactive glass-ceramic

Neda Ghaebi Panah[1], Rob Atkin[2] and Tim Sercombe[*1]

[1]School of Engineering, University of Western Australia, 35 Stirling Highway, Perth, WA 6009, Australia
[2]School of Molecular Sciences, University of Western Australia, 35 Stirling Highway, Perth, WA 6009, Australia

[*]Tim.sercombe@uwa.edu.au

Abstract. Mesoporous bioactive glass ceramic in the system of $60SiO_2$-$36CaO$-$4P_2O_5$ (mol.%) named 58S was synthesized using the sol-gel process. The effect of different drying methods (oven-drying, freeze drying, and vacuum drying) on the structure of 58S bioactive glass ceramic was investigated. Samples were characterized by scanning electron microscopy (SEM) and nitrogen sorption porosimetry (BET). SEM images of the samples after drying and then calcination indicated crystals formed in the 58S glass. The nitrogen isotherm of all samples calcinated at 600 °C revealed a mesoporous structure for 58S glass-ceramics. BET surface area of freeze-dried sample after calcination was greater than that of the oven and vacuum dried specimens. Drying methods had significant effect on morphology, surface area, and shape of pores.

1. Introduction

Bioactive glasses were the first group of synthetic biomaterials which showed a bone-bonding ability [1]. If implanted in the body, a bone-like apatite phase forms on their surface, which forms a strong bond with bone. Apart from bioactivity, they are also biocompatible and biodegradable [1], [2]. Bioactive glasses can be synthesized either by melting or sol-gel process [2].
Sol-gel is a wet-chemical process in which inorganic materials as well as organic-inorganic hybrids are synthesized from liquid sources at low temperatures. This method is particularly attractive for preparing glasses with a high tendency to crystallize and relatively high melting temperature [3].

Other advantages of the sol-gel process over the melting technique for making bioactive glasses are higher bioactivity, higher surface area, wider range of bioactive compositions, and interconnectivity between pores [1], [4]. Through the sol-gel, a solid porous network in a liquid forms due to the hydrolysis of alkoxides. Next polycondensation creates a gel. In order to remove the liquid phase which fills the porosity of the solid network, a drying step is required. Drying the gel by heating at low temperature results in a xerogel, while freeze drying makes cryogel. The methods of liquid extraction within the structure can induce different levels of capillary stresses on the solid network [5] which may change the surface structure. The aim of this study is to compare the surface structure of a gel which is dried using oven drying, freeze drying, and vacuum drying with using the drying mechanisms of evaporation, sublimation, and boiling, respectively.

2. Experimental procedure and characterization methods

The chemical composition of 58S glass, $60SiO_2$-$36CaO$-$4P_2O_5$ (mol.%), was chosen based on the sol-gel calcium silicate bioglass [1]. To synthesize 50g of 58S bioactive glass, 100.940 g Tetraethyl orthosilicate (TEOS, product No. 86578 from Sigma Aldrich) was added into the mixture of distilled water, ethanol absolute (product No. 214 from Ajax Finechem Chemicals), and nitric acid 69% (HNO_3, product No. 101799 form Merck) with the molar ratio of (HNO_3+H_2O+EtOH)/(TEOS+TEP)=8 in TEOS:H_2O:EtOH:HNO_3=0.48:4:0.34:0.02. The pH of the solution was adjust to 0.84. The solution was magnetically stirred for 1h at room temperature to hydrolyze TEOS. Next, 11.767 g triethyl phosphate (TEP, No. 30505 from BDH Chemicals Ltd Poole England) and 68.654 g calcium nitrate tetrahydrate ($Ca(NO_3)_2.4H_2O$, product No. CA015 from Chem-supply) were added into the solution sequentially, with an interval of 1h between each addition. The final solution (clear sol) was stirred for an additional 1 hour, then poured into a glass petri dish, and stored at room temperature until they formed a clear gel. The clear gel was dried using different methods of oven-drying, freeze drying, and vacuum drying, named O-58S, F-58S, and V-58S, respectively. In the case of oven-drying, the gel was aged at 70 °C for 24h and then dried at 120 °C for 24h. Vacuum drying was employed for 36h. In the case of freeze drying, the gel was placed into the liquid nitrogen and then freeze drying was carried out for 24h on a Labconco FreeZone Plus 6. The dried gels were subsequently calcined in air by heating them at 2°C/min to 600 °C and holding for 3h. After calcination, the glass-ceramics were ground and until are passed through a 74 microns sieve.

The structure and morphology of the 58S bioactive glass-ceramics were observed using scanning electron microscopy on a Zeiss 1555 VP-FESEM working at 5 kV, 5.2-5.6 WD, with an In-Lens detector. The samples were Pt-coated before imaging. Nitrogen adsorption-desorption isotherm measurements were accomplished on Micromeritics ASAP 2020 at 77 K using a relative pressure from 0 to 0.99 to determine the specific surface area, pore size, and pore volume.

3. Results and discussion

The morphology and structure of 58S bioactive glass-ceramics after different drying methods are shown in Figure 1. Comparing the SEM images reveals that the O-58S glass-ceramic contains an agglomeration of particles, the F-58S is cracked and the V-58S has a highly rough surface. It is apparent from Figure 1(a) and (b) that particles have become agglomerated through the thermal drying. In contrast, freeze-drying creates a different morphology with some cracks evident within the particles, Figure 1(c) and (d). These cracks may have occurred during the sublimation process. In other words, after freezing, sublimation of the solvent may lead to cracking. At room temperature, the gel is able to flow into the space created by the removed of the solvent and therefore cracking does not occur. The same flow is not possible when the gel is frozen and therefore cracking occurs. Finally SEM images of the vacuum dried gel, Figure 1(e) and (f) show a rough, porous surface which may be attributed to the way that the solvent molecules leave the gel under vacuum.

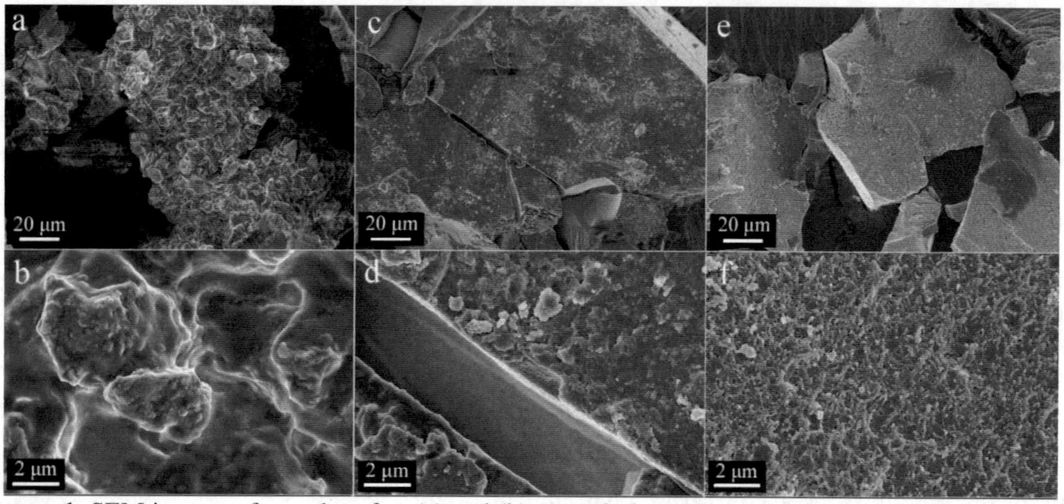

Figure 1. SEM images of samples after (a) and (b) oven drying, (c) and (d) freeze drying, and (e) and (f) vacuum drying. Images (b), (d) and (f) are higher magnification images of (a), (c) and (e), respectively.

Figure 2 shows the SEM images of O-58S, F-58S, and V-58S after calcination at 600 °C for 3h. Comparing the morphologies of the material after calcination indicates that the drying method causes a significant change in the surface structure. The O-58S was partially crystallized (Figure 2(a) and (b)) with flake like crystals. For the F-58S sample, there were areas that look similar to the O-58S but at some regions were different such as shown in Figure 2(c) and (d). Vacuum drying, Figure 2(e) and (f), results in a similar flake like crystalline morphology to oven drying. However, the crystals of V-58S are more interlocked and smaller. From these results, it is apparent that the method of solvent extraction from the gel affects the surface porosity size and shape as well as agglomeration of particles.

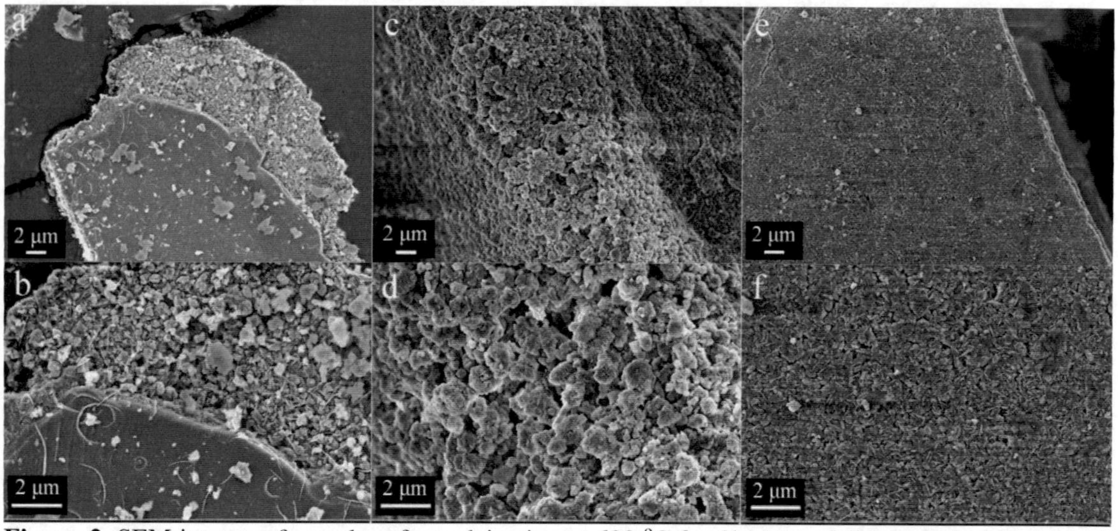

Figure 2. SEM images of samples after calcination at 600 °C for 3h. (a) and (b) oven drying, (c) and (d) freeze drying, and (e) and (f) vacuum drying. Images (b), (d) and (f) are higher magnification images of (a), (c) and (e), respectively.

Figure 3 shows the nitrogen sorption isotherms of O-58S, F-58S, and V-58S after calcination at 600 °C for 3h. The nitrogen sorption isotherms of all samples are identified as type IV according to IUPAC classification, with a hysteresis loop which is representative of mesoporous materials [6]. Table 1 summarizes BET surface area, pore volume, and pore size of O-58S, F-58S, and V-58S after calcination. Calcined F-58S had the highest BET surface area with lowest pore size amongst the samples. Based on the results of nitrogen sorption isotherms, there is a significant difference amongst the BET surface area of samples, although all of them represented mesoporous structure.

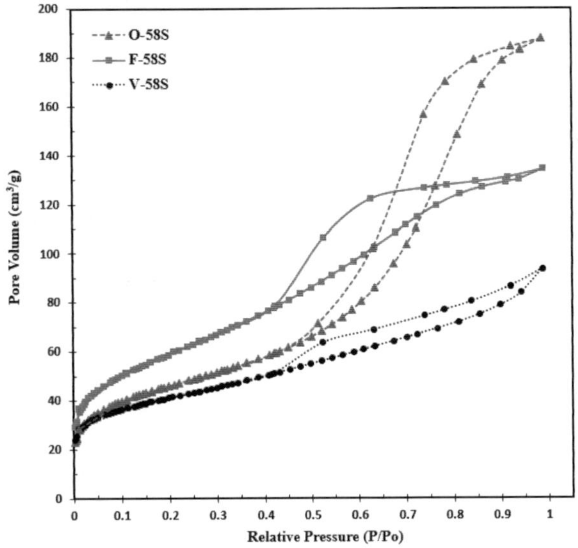

Figure 3. Nitrogen sorption isotherms of O-58S, F-58S, and V-58S calcinated at 600 °C for 3h.

Table 1. BET surface area, pore volume, and pore size of O-58S, F-58S, and V-58S calcinated at 600 °C for 3h.

	O-58S after calcination	F-58S after calcination	V-58S after calcination
BET surface area (m^2/g)	161	209	140
Pore volume (cm^3/g)	0.29	0.21	0.15
Pore size (nm)	7.20	3.97	4.12

4. Conclusions

Mesoporous bioactive glass ceramics in the system of $60SiO_2$-$36CaO$-$4P_2O_5$ (mol.%) (i.e. 58S) have been synthesized using the sol-gel process. The effect of different drying methods (oven-drying, freeze drying, and vacuum drying) on the structure of 58S bioactive glass ceramic was investigated. SEM images of the samples after drying revealed an agglomeration of the particles for oven-dried sample, cracks in the freeze-dried material, and a highly rough surface on the vacuum dried particles. After calcination, both the oven and vacuum dried material were partially crystallized with flake like crystals, while freeze-drying produced a different crystalline morphology. The nitrogen isotherms of all samples calcinated at 600 °C indicated that a mesoporous structure formed. The BET surface area of the freeze-dried materials after calcination was greater than after either oven or vacuum drying. Hence, it is apparent that the method of solvent extraction from the gel has significant effect on morphology and surface properties for the resultant bioactive glass ceramic.

Acknowledgments

Australian Government Research Training Program PhD Scholarship from The University of Western Australia is gratefully acknowledged. The authors also acknowledge the facilities, and the scientific and technical assistance of the Australian Microscopy & Microanalysis Research Facility at the Centre for Microscopy, Characterisation & Analysis, The University of Western Australia, a facility funded by the University, State and Commonwealth Governments.

IOP Conf. Series: Materials Science and Engineering **856** (2020) 012003 doi:10.1088/1757-899X/856/1/012003

References

[1] S. Kargozar, M. Montazerian, S. Hamzehlou, H.-W. Kim, and F. Baino 2018 "Mesoporous bioactive glasses: Promising platforms for antibacterial strategies" *Acta Biomater.* **81** 1–19.

[2] F. Baino 2018 "Bioactive glasses – When glass science and technology meet regenerative medicine" *Ceram. Int.* **44** 14953–66.

[3] K. Kajihara 2013 "Recent advances in sol–gel synthesis of monolithic silica and silica-based glasses" *J. Asian Ceram. Soc.* **1** 121–33.

[4] S. Arabyazdi, A. Yazdanpanah, A. Ansari Hamedani, A. Ramedani, and F. Moztarzadeh 2019 "Synthesis and characterization of CaO-P_2O_5-SiO_2-Li_2O-Fe_2O_3 bioactive glasses: The effect of Li_2O-Fe_2O_3 content on the structure and in-vitro bioactivity" *J. Non. Cryst. Solids* **503–504** 139–50.

[5] J. Phalippou 2002 "From gel to glass" *Comptes Rendus Chim.* **5** 855–863.

[6] S. Kunjalukkal Padmanabhan, E. Ul Haq, and A. Licciulli 2016 "Synthesis of silica cryogel-glass fiber blanket by vacuum drying" *Ceram. Int.* **42** 7216–22.

Theoretical exploration on free radical scavenging mechanism of curcumin analogues in water solvent

L S P Boli[1,2], N D Aisyah[2], V Khoirunisa[1,3], F Rusydi[2,4], H Rachmawati[5], H K Dipojono[1]*

[1]Department of Engineering Physics, Faculty of Industrial Engineering, Institut Teknologi Bandung, Bandung 40132, Indonesia
[2]Research Center for Quantum Engineering Design, Universitas Airlangga, Surabaya 60115, Indonesia
[3]Engineering Physics Program, Institut Teknologi Sumatera, Lampung Selatan 35365, Indonesia
[4]Department of Physics, Universitas Airlangga, Surabaya 60115, Indonesia
[5]School of Pharmacy, Institut Teknologi Bandung, Bandung 40132, Indonesia
Department of Physics

*dipojono@gmail.com

Abstract. One way to improve curcumin's possibility in preventing lipid peroxidation is through modification of carbon linker, β-diketone group and aromatic rings. However, there remains a need to explore the role of carbon linker on the free radical scavenging mechanisms of curcumin. This study uses density functional theory to explore two descriptors of free radical scavenging mechanisms, which are bond dissociation enthalpy (BDE) and adiabatic ionization potential (AIP) of curcumin and its analogues. Five analogues with different amount of carbon atoms in the linker in the presence / absence of β-diketone group are chosen. Our result shows that decreasing the amount of carbon atom from 7 to 5 atoms in the linker decreases the BDE at β-diketone group of the curcumin analogue. Moreover, increasing the amount of carbon atoms in the linker decreases the AIP of the analogues.

1. Introduction

Research attention on curcumin is increasing due to its antioxidant activities that offer extensive applications in the health-related fields. Curcumin ($C_{21}H_{20}O_6$) is a yellow compound isolated from the rhizome of Curcuma longa. In physiological conditions, curcumin exists in keto and enol tautomers [1,2]. Research about curcumin has been carried to understand antioxidant activity of curcumin which include its free radicals scavenging activity [3] and lipid peroxidation reduction [4]. This activity enables curcumin for application in treating diseases such as Alzheimer's disease, β-Thalassemia, kidney injury, skin disorders, and non-alcoholic fatty liver disease [5–9].

In performing its antioxidant activity, curcumin scavenges free radicals through different scavenging mechanisms. The mechanisms are hydrogen atom transfer (HAT), single electron transfer followed by proton transfer (SETPT), and sequential proton loss electron transfer (SPLET). The mechanisms occur depending on the environment in which curcumin exists and the radicals that react toward curcumin. For example, experiment by Jovanovic showed that in pH 3 to 7 where curcumin keto dominates, HAT mechanism rules the scavenging activity. Meanwhile, in pH above 8 where curcumin enol dominates, SET mechanism governs the activity [10]. Furthermore, theoretical calculation by Galano

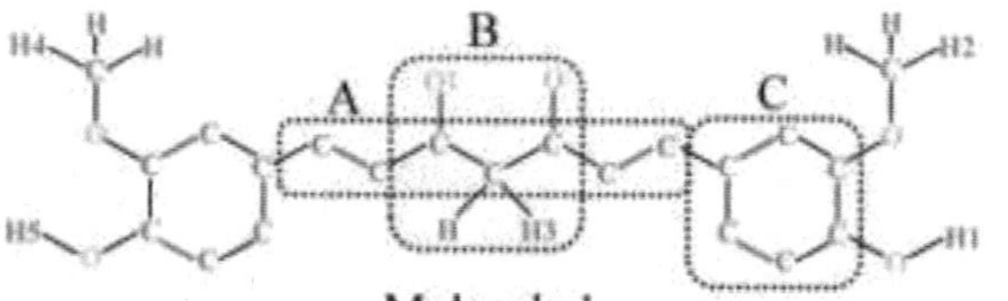

Molecule 1

Molecule	Part A	Part B
2	7 C atoms	H3 atom is attached to O1 atom
3	5 C atoms	Same as Molecule 2
4	9 C atoms	Same as Molecule 2
5	5 C atoms	Monocarbonyl
6	7 C atoms	Same as Molecule 5
7	9 C atoms	Same as Molecule 5

Scheme 1. Structure of curcumin and its modified structures. Part **A**, **B**, and **C** denote carbon linker, β-diketone group and aromatic ring.

et al. displayed that HAT and SPLET mechanisms are dominant when curcumin reacts with methoxyl radical ($\bullet OCH_3$) and trichloromethylperoxyl radical ($\bullet O_2CCl_3$) respectively [3].
Curcumin modification is an attempt to improve the free radical scavenging activity. Geometry of curcumin contains carbon linker, β-diketone group and two aromatic rings (Scheme 1). Previous studies have reported that the modification of β-diketone group affects the scavenging mechanism. Mary et al. modified curcumin by forming a metal complex between β-diketone group and certain metal ion and found that the complex enhances HAT and SET mechanism [11]. Li et al. reported that modified curcumin with shorter carbon linker but without β-diketone group is able to scavenge 1,1-diphenyl-2-picryl-hydrazyl (DPPH) radical through the SET mechanism experimentally [12]. Considering the role of the β-diketone group and the carbon linker, further exploration of the linker effect, in the presence and absence of the group, on scavenging mechanisms is necessary.
This paper explores the role of carbon linker on the free radical scavenging mechanism of curcumin by utilizing density functional theory (DFT) calculation to calculate two thermochemical parameters of curcumin analogues. The parameters are bond dissociation enthalpy (BDE) and adiabatic ionization potential (AIP). BDE is measured as the energy releases when an X-H bond of the curcumin is breaking prior to hydrogen transfer, while AIP is measured as the energy required to remove an electron from the curcumin prior to electron transfer. Both BDE and AIP are descriptors for HAT and SET mechanisms respectively.

2. Computational Detail

This research is conducted to explore hydrogen atom transfer and electron transfer, as modelled by (1) and (2), in two stages. First is performing DFT calculation to obtain the geometry of all molecules at ground state. The molecules involved are neutral, radical, and radical ion. The neutral molecules consist of two curcumin tautomers: keto (Molecule 1, with 7 carbon atoms in the linker) and enol (Molecule 2, also with seven carbon atom in the linker), two analogues with 5 and 9 carbon atoms in the linker with β-diketone group (Molecule 3 and 4) and three analogues with 5, 7 and 9 carbon atoms in the linker without β-diketone group (Molecule 5, 6 and 7), as illustrated in scheme 1. The radicals are constructed through hydrogen abstraction at certain X-Hn site (n = 1, 2, 3, 4, and 5) of the neutral molecules. Meanwhile, the radical ions are constructed by applying +1 charge to the neutral molecules. To obtain the geometry of the molecules at ground state, geometry optimization coupled with frequency calculation are performed using M06-2X/6-311++G(d,p) for the neutral molecules and UM06-2X/6-311++G(d,p) for the radical and the radical ion. This step will produce electronic energy and thermal correction to enthalpy of each molecules. Water solvent is included to the calculation by using polarized continuum model (PCM) calculation. All calculation is performed at 298.15 K and 1 atm using Gaussian 09 software [13].
Second is calculating BDE and AIP based on the result of the DFT calculation. The BDE and AIP are calculated according to these models and are presented in (3) and (4) respectively.

IOP Conf. Series: Materials Science and Engineering **856** (2020) 012004 doi:10.1088/1757-899X/856/1/012004

Table 1. Bond dissociation enthalpy of the seven molecules in water solvent.

Molecule	Abstraction Site	BDE (kcal/mol) Calc.	Ref.[a]	Ref.[b]
1	O-H1	84.3	78.3&78.5	82.3
	C-H2	97.9		
	C-H3	88.7	92.7	87.4
2	O-H1	84.0	77.5	80.2
	C-H2	96.2		
	O-H3	106.8		
	C-H4	97.8		
	O-H5	83.6	76.0	
3	O-H1	84.0		
	C-H2	97.7		
	O-H3	89.6		
	C-H4	98.2		
	O-H5	84.3		

Molecule	Abstraction Site	BDE (Calc., kcal/mol)
4	O-H1	83.0
	C-H2	97.6
	O-H3	104.6
	C-H4	97.7
	O-H5	83.2
5	O-H1	84.1
	C-H2	97.8
6	O-H1	83.3
	C-H2	97.7
7	O-H1	82.8
	C-H2	96.1

[a] Computed using B3LYP/6-311+G(2d,2p)//AM1 calculations. [14]
[b] Computed using (RO) B3LYP/6-311+G (2d,2p)//AM1/AM1 calculations. [15]

For this calculation, enthalpy of hydrogen (H(H·)) in water solvent is taken from the value used by szelag et al. [16].

$$AXH \rightarrow AX\cdot + H\cdot \quad (1)$$

$$AXH \rightarrow AXH\cdot^{+} + e^{-} \quad (2)$$

$$BDE = H(AX\cdot) + H(H\cdot) - H(AXH) \quad (3)$$

$$AIP = H(AXH\cdot^{+}) - H(AXH) \quad (4)$$

Where AXH, AX· and H· represents the 7 molecules, the radical form of the 7 molecules, and a hydrogen atom, respectively.

3. Results and Discussion

3.1. Bond dissociation enthalpy of curcumin analogues

Table 1 lists the chosen abstraction site and the results of the calculated BDE at each abstraction site. Molecule 1, 5, 6 and 7 have symmetrical shape, thus only O-H1 and C-H2 attached to one aromatic ring are chosen as the abstraction sites. C-H3 site in Molecule 1 is chosen as an extra abstraction site since it has been proposed by Jovanovic et al. as curcumin's possible abstraction site [9]. Molecule 2, 3 and 4 are not symmetrical, hence there are five abstraction sites chosen. Our results show that in molecule 1, BDE of O-H1 < C-H3 < C-H2. In Molecule 2, BDE of O-H1 ≈ O-H5 < C-H2 < C-H4 < O-H3. Comparing the BDEs in Molecule 1 and 2, it is concluded that O-H sites in aromatic ring always have the lowest BDEs and their values are similar. This is in agreement with the result reported by Barzegar et al. and Sun et al. [14,15].

Changing the amount of carbon atoms in the linker does not affect the BDE at aromatic ring but it does at the β-diketone group. Molecule 3, 2 and 4 that have 5, 7 and 9 carbon atoms in the linker respectively in the presence of the β-diketone group, show similar BDE values at O-H and C-H sites of aromatic ring. Molecule 5, 6, and 7 that have 5, 7 and 9 carbon atoms in the linker respectively in

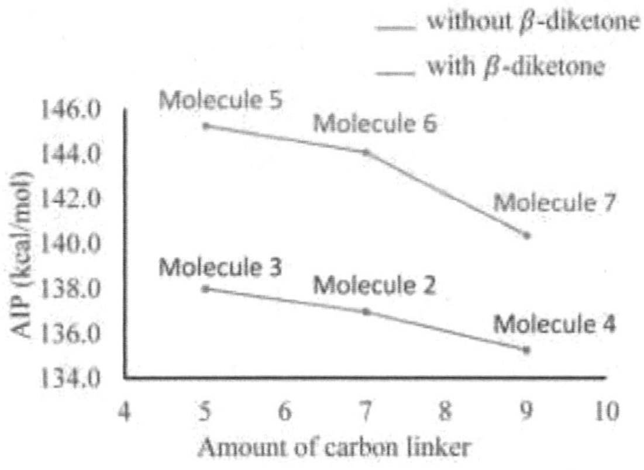

Figure 1. Plot of the AIP of the seven molecules.

Table 2. The HOMO energy of the seven molecules.

Molecule	E_{HOMO} (eV)
1	-7.36
2	-7.05
3	-7.33
4	-7.11
5	-7.21
6	-7.08
7	-6.92

the absence of the β-diketone group also show similar BDE values at sites in aromatic ring. Regardless of the amount of carbon linker and the presence / absence of β-diketone group, the BDE at O-H site in aromatic ring remains the lowest among all abstraction sites. Different result is observed at O-H3 sites of β-diketone group in Molecule 2, 3 and 4. Molecule 3 that has 5 carbon atoms in the linker has 17.2 kcal/mol lower BDE value than Molecule 2 that has 7 carbon atoms in the linker. Meanwhile Molecule 4 that has 9 carbon atoms in the linker has only 2.2 kcal/mol lower BDE value than Molecule 2.

This results suggest two important things. First, since low BDE relates to an easy HAT, the seven molecules probably can transfer a hydrogen atom from O-H site at aromatic ring easily. Thus, it is possible for curcumin and its analogues to participate in scavenging free radicals through HAT mechanism. Furthermore, the C-H3 site in Molecule 1 also has to be considered as a possible hydrogen atom transfer site, which has been supported by Jovanovic et al. [9]. Second, changing amount of carbon atom in the linker in the presence of β-diketone group is important since the linker has shown to affect the BDE at the β-diketone group.

3.2. Adiabatic ionization potential of curcumin analogues

Figure 1 plots the calculated AIP of two groups of curcumin analogues, with and without β-diketone group. The figure shows that increasing the amount of carbon atoms from 5 to 7 to 9 reduces the AIP values in both groups with an average of 3.8 kcal/mol. Overall, molecules with β-diketone group has lower AIP values. The molecules with lower AIP values can facilitate SET, the first step of SETPT mechanism, easier [16]. Thus, our result implies that increasing of the amount of carbon linker in the presence of β-diketone group improves the possibility of the molecules to scavenge free radicals through SETPT mechanism.

Table 2 lists the results of highest occupied molecular orbital (HOMO) energy of the seven molecules. Generally, increasing the amount of carbon linker increases the HOMO energy, which could be an indication of an easy electron transfer [17]. Thus, the HOMO results confirm the AIP trend analyzed above.

4. Conclusion

This study uses two descriptors, BDE and AIP, to explore the role of carbon linker on the free radical scavenging mechanisms of curcumin. Decreasing the amount of carbon atom in the linker lowers the BDE value at β-diketone group by 17.2 kcal/mol. This implies that shorter carbon linker probably can increase the possibility of curcumin to scavenge free radical through HAT mechanism at β-diketone group. On the other hand, increasing the amount of carbon atom in the linker reduces the AIP value by 3.8 kcal/mol on average. This implies that longer carbon linker probably can increase the possibility of curcumin to involve in SETPT mechanism.

Acknowledgments
This work is supported by Directorate of Higher Education, Ministry of Research Technology and Higher Education (RISTEKDIKTI), Republic of Indonesia under grant scheme "Penelitian Dasar Unggulan Perguruan Tinggi 2019". LSPB appreciates support from Lembaga Pengelola Dana Pendidikan (LPDP) for the doctoral scholarship.

References

[1] Tønnesen H H, Karlsen J, Mostad A, Samuelsson B, Enzell C R and Berg J-E 1982 *Acta Chem. Scand. B* **36** 475–9
[2] Sharma R A, Gescher A J and Steward W P 2005 *Eur. J. Cancer* **41** 1955–68
[3] Galano A, Álvarez-Diduk R, Ramírez-Silva M T, Alarcón-Ángeles G and Rojas-Hernández A 2009 *Chem. Phys.* **363** 13–23
[4] Chaniad P, Morales N P, Rojsitthisak P and Luechapudiporn R 2018 *J. Food Biochem.* **42** 1–9
[5] Bisceglia F, Seghetti F, Serra M, Zusso M, Gervasoni S, Verga L, Vistoli G, Lanni C, Catanzaro M, De Lorenzi E and Belluti F 2019 *ACS Chem Neurosci.* **10** 1420-1433
[6] Nasseri E, Mohammadi E, Tamaddoni A, Qujeq D, Zayeri F and Zand H 2017 *Ann. Nutr. Metab.* **71** 136–44
[7] Abdel-Magied N and Elkady A A 2019 *Exp. Mol. Pathol.* **111** 104299
[8] Laura V, Mattia F, Roberta G, Federico I, Emi D, Chiara T, Luca B and Elena C 2019 *Nutrients* **11** 2169
[9] Mirhafez S R, Farimani A R, Gholami A, Hooshmand E, Tavallaie S and Nobakht M. Gh B F 2019 *Drug Metab. Pers. Ther.* **34** 20180040
[10] Jovanovic S V, Steenken S, Boone C W and Simic M G 1999 *J. Am. Chem. Soc.* 9677–81
[11] Mary C P V, Vijayakumar S and Shankar R 2018 *J. Mol. Graph. Model.* **79** 1–14
[12] Li Q, Chen J, Luo S, Xu J, Huang Q and Liu T 2015 Eur. *J. Med. Chem.* **93** 461–9
[13] Gaussian 09, Revision D.01, M. J. Frisch, G. W. Trucks, H. B. Schlegel, G. E. Scuseria, M. A. Robb, J. R. Cheeseman, G. Scalmani, V. Barone, B. Mennucci, G. A. Petersson, Nakatsuji, M. Caricato, X. Li, H. P. Hratchian, A. F. Izmaylov, J. Bloino, G. Zheng, J. L. Sonnenberg, M. Hada, M. Ehara, K. Toyota, R. Fukuda, J. Hasegawa, M. Ishida, T. Nakajima, Y. Honda, O. Kitao, H. Nakai, T. Vreven, J. A. Montgomery, Jr., J. E. Peralta, F. Ogliaro, M. Bearpark, J. J. Heyd, E. Brothers, K. N. Kudin, V. N. Staroverov, T. Keith, R. Kobayashi, J. Normand, K. Raghavachari, A. Rendell, J. C. Burant, S. S. Iyengar, J. Tomasi, M. Cossi, N. Rega, J. M. Millam, M. Klene, J. E. Knox, J. B. Cross, V. Bakken, C. Adamo, J. Jaramillo, R. Gomperts, R. E. Stratmann, O. Yazyev, A. J. Austin, R. Cammi, C. Pomelli, J. W. Ochterski, R. L. Martin, K. Morokuma, V. G. Zakrzewski, G. A. Voth, P. Salvador, J. J. Dannenberg, S. Dapprich, A. D. Daniels, O. Farkas, J. B. Foresman, J. V. Ortiz, J. Cioslowski, and D. J. Fox, Gaussian, Inc., Wallingford CT, 2013.
[14] Abolfazl Barzegar 2012 *Food Chem.* **135** 1369-1376
[15] You-Min Sun, Hong-Yu Zhang, De-Zhan Chen, and Cheng-Bu Liu 2002 *Org. Lett.* **4** 2909-2911
[16] Małgorzata Szeląg, Alicja Urbaniak, Hans A.R. Bluyssen 2015 *Open Chem.* **13** 17-31
[17] G. Mazzone et al., 2015 *Comput. Theoret. Chem.* **1077** 39-47

IOP Conf. Series: Materials Science and Engineering **856** (2020) 012005 doi:10.1088/1757-899X/856/1/012005

Theoretical investigation on electron transfer-based antioxidant activity of melinjo resveratrol

V Khoirunisa[1,2,3], R N Fadilla[3], L S P Boli[1,3], F Rusydi[3,4], H Rachmawati[5] and H K Dipojono[1]*

[1]Department of Engineering Physics, Faculty of Industrial Engineering, Institut Teknologi Bandung, Bandung 40132, Indonesia
[2]Engineering Physics Program, Institut Teknologi Sumatera, Lampung Selatan 35365, Indonesia
[3]Research Center for Quantum Engineering Design, Universitas Airlangga, Surabaya 60115, Indonesia
[4]Department of Physics, Faculty of Science and Technology, Universitas Airlangga, Surabaya 60115, Indonesia
[5]School of Pharmacy, Institut Teknologi Bandung, Bandung 40132, Indonesia

* dipojono@gmail.com

Abstract. We employ density functional calculations to study the electron transfer-based antioxidant activity of resveratrol derivatives from melinjo. The derivatives are monomer resveratrol which is trans-resveratrol, and dimer resveratrol which are gnetin C, gnetin L, and gnemonoside C. The results show that gnetin C has the lowest adiabatic ionization potential both in gas phase and water environment. Furthermore, our results demonstrate that dimer resveratrol is consistent to have better electron transfer ability than its monomer. This may due to the destabilization of the high occupied molecular orbital energy of resveratrol. However, the presence of glucoside in gnemonoside C does not affect the electron transfer-based antioxidant activity of dimer resveratrol.

1. Introduction

Resveratrol has gained great attention due to its prominent antioxidant properties. Through various *in vitro* assays, Gulcin has proven the resveratrol's excellent antioxidant activity compared to some standard antioxidant viz. trolox, BHA, BHT, and α-tocopherol [1]. Furthermore, some studies also support that resveratrol exhibits antioxidant effect in some physiological systems [2,3,4]. The antioxidant properties are linked to promote various health benefits such as protecting the brain [5] and lowering blood pressure [6]. Hence, resveratrol become a growing interest to be drug reference molecule.

Resveratrol is a phenolic antioxidant that can deactivate radicals by two possible mechanisms, hydrogen atom transfer (HAT) and single electron transfer mechanism (SET) [7]. HAT is a direct mechanism, while SET is a stepwise mechanism. Both of them have the same net reaction in which antioxidant donate one of its hydrogen atoms to radical at the end of reaction. Even though HAT

IOP Conf. Series: Materials Science and Engineering **856** (2020) 012005 doi:10.1088/1757-899X/856/1/012005

mechanism is predominant mechanism, it is essential to investigate SET mechanism to explore antioxidant behaviour of molecules.

One of resveratrol derivatives sources is melinjo seed. The seed contains resveratrol mostly in the form of dimers [8] which are gnetin and gnemonoside. Gnetin is the simplest resveratrol dimer, and gnemonoside is a resveratrol dimer with additional glucoside. Experimental study performed by Kato et al. suggested that antioxidant activity of melinjo seed may associate with electron-donating ability of resveratrol dimer [9]. However, to the best of our knowledge, no further investigation is reported to confirm electron-donating ability of resveratrol dimer.

This paper address the relation between antioxidant activity of melinjo resveratrol and its structure based on SET mechanism; that contributes to a better understanding on the electron-donating ability of resveratrol dimer. Four chosen resveratrol derivatives are trans-resveratrol, gnetin C, gnetin L, and gnemonoside C (figure 1). The parameter used to examine the antioxidant activity of the four resveratrol derivatives is adiabatic ionization potential (AIP). This parameter is determined by performing density functional calculations.

Finally, the insightful discussion about the AIP values contribute to a better understanding on the electron-donating ability of resveratrol dimer.

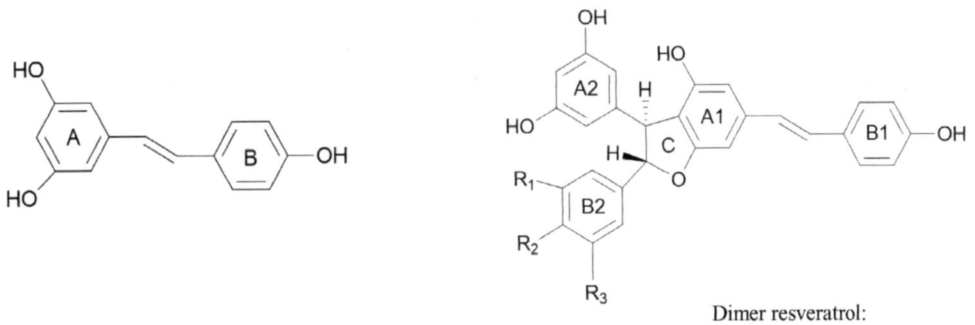

Monomer Resveratrol:
A. Trans-resveratrol

Dimer resveratrol:
B. Gnetin C: R_1=H; R_2=OH; R_3=H
C. Gnetin L: R_1=OH; R_2=H; R_3 3=OCH3
D. Gnemonoside C: R_1=H; R_2=O-glucoside; R_3=H

Figure 1. Four resveratrol derivatives from melinjo seed.

2. Computational Model

We define our reaction model based on SET mechanism as it is shown in equation 1:

$$XH \rightarrow XH^{+} + e^{-}, \tag{1}$$

where XH and XH^{+} represent the parent molecules and the respective cation radical generated after electron transfer. We perform DFT calculations by using Gaussian 09 software package [10]. We use Becke three parameter exchange and Lee–Yang–Parr correlation (B3LYP) and 6-31++G(d,p) basis sets for geometry optimization of each resveratrol and respective cation structure in gas phase and water environment. The water solvation is considered using polarizable continuum model. We also perform vibrational frequency calculation to confirm that there is no imaginary frequency in the ground state structure, and to give thermal correction at 298.15 K. AIP is determined according to the formula,

$$AIP = (H_{XH^{+}} + H_{e-}) - H_{XH}, \tag{2}$$

where H_{e-} is enthalpy of the electron (0.00119787 hartree in gas phase and -0.0492478 hartree in water) [11].

3. Results and Discussion

IOP Conf. Series: Materials Science and Engineering **856** (2020) 012005 doi:10.1088/1757-899X/856/1/012005

Table 1 shows the AIP values of the resveratrol derivatives in gas phase and water environment. It shows that the AIP values of trans-resveratrol in gas phase and water environment are 160.7 and 93.7 kcal/mol, respectively. These values are close to the results of previous works [12,13]. Therefore, our model is reasonable for studying this case.

The values of AIP in water environment is lower than that in gas phase. We suspect that it is due to the polarity of the water that is higher than the gas phase. This result agrees with previous work done by Mikulski et al. [12]. They demonstrated that the AIP of molecules decreases with the increasing environment polarity. The lower AIP value in water environment indicates that the resveratrol derivatives are easier to transfer electron in water environment than in gas phase. The polar environment may facilitate the electron donation that is important for radical scavenging activity of the resveratrol derivatives [13].

We also find that all resveratrol dimers have lower AIP values than its monomer, trans-resveratrol. It implies that dimer structure of resveratrol has better electron-donating ability than the monomer. The sequence of AIP values of the resveratrol derivatives is as follows: **gnetin C < gnetin L or gnemonoside C < trans-resveratrol**. Thus, we propose gnetin C to be the most potential electron transfer-based antioxidant among the four resveratrol derivatives due to its highest electron-donating ability.

Table 1. AIP values of four resveratrol derivatives (see figure 1) in gas phase and in water environment in terms of enthalpy.

	AIP (kcal/mol)	
Resveratrol derivatives	Gas phase	Water
Trans-resveratrol	160.7	93.7
Gnetin C	152.8	92.3
Gnetin L	154.0	92.4
Gnemonoside C	154.0	N/A

Table 2 presents the HOMO energy of four resveratrol derivatives. We find that the stability of the HOMO energy corresponds to the AIP value. The more stable the HOMO energy, the higher the AIP value is. This relation is plotted in figure 2. By considering this relation, we may infer that the increasing electron transfer ability is associated with the destabilization of the HOMO energy. In this case, the destabilization in gas phase that is up to 0.18 eV corresponds to the decrement of AIP value by 7.9 kcal/mol.

Table 2. HOMO energy of four resveratrol derivatives in gas phase and in water environment.

	HOMO energy (eV)	
Resveratrol derivatives	Gas phase	Water
Trans-resveratrol	-5.63	-5.67
Gnetin C	-5.45	-5.60
Gnetin L	-5.51	-5.61
Gnemonoside C	-5.53	N/A

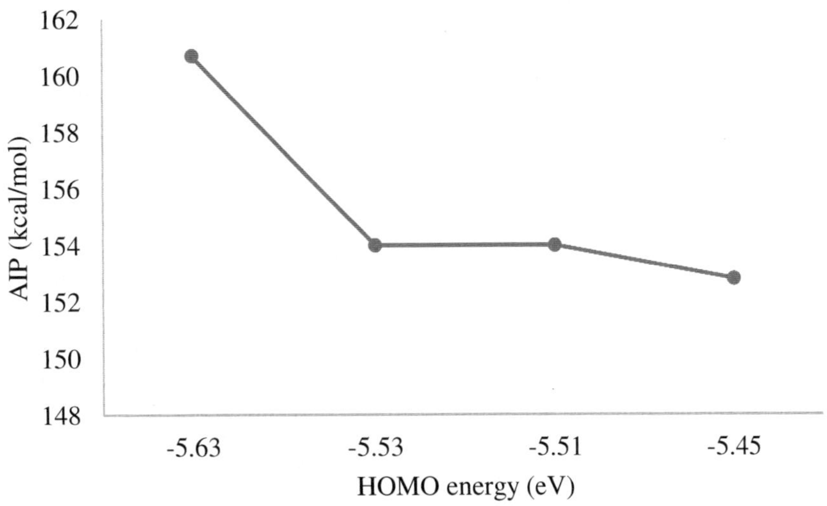

Figure 2. AIP values against HOMO energy of four resveratrol derivatives in gas phase. The HOMO energy is more stable towards the left side of the *x*-axis.

We also note that gnemonoside C has relatively close HOMO energy to gnetin C even though they are differed by the presence of glucoside. Based on our calculation results, the presence of glucoside slightly stabilizes the HOMO energy that leads to small changes of AIP value. Therefore, in this case, the presence of glucoside is not necessarily considered.

4. Conclusion

We have explored the electron-based antioxidant activity of four resveratrol derivatives: trans-resveratrol, gnetin C, gnetin L, and gnemonoside C. Trend of the AIP values has been found to be correlated with the trend of the HOMO energy. We suggest that the HOMO energy destabilization can enhance the antioxidant activity of resveratrol. Based on our finding, we propose that antioxidant activity of four melinjo resveratrols follows the order: **resveratrol dimer > resveratrol dimer with glucoside > resveratrol monomer**. To further our study, we are planning to investigate the reaction pathways of electron-donating precess from resveratrol dimer to radicals.

Acknowledgments

This work is supported by Directorate of Higher Education, Ministry of Research Technology and Higher Education (RISTEKDIKTI), Republic of Indonesia under grant scheme "Penelitian Dasar Unggulan Perguruan Tinggi 2019". VK appreciates support from Lembaga Pengelola Dana Pendidikan (LPDP) for the doctoral scholarship.

References

[1] Gülçin I 2010 *Innov. Food Sci. Emerg. Technol.* **11** 210
[2] Liu Y and Liu G 2004 *Biochem Pharmacol.* **67** 777
[3] Olas B, Wachowicz B, Nowak P, Stochmal A, Oleszek W, Glowacki R, Bald E 2008 *Cell Biol. Toxicol.* **24** 331
[4] Momken I, Stevens L, Bergouignan A, Desplanches D, Rudwill F, Chery I, Zahariev A, Zahn S, Stein PT, Sebedio JL, Pujos-Guillot E, Falempin M, Simon C, Coxam V, Andrianjafiniony T, Gauquelin-Koch G, Picquet F, and Blanc S 2011 *The FASEB Journal* **25** 3646
[5] Braidy N, Jugder BE, Poljak A, Jayasena T, Mansour H, Nabavi SM, Sachdev P, and Grant R

IOP Conf. Series: Materials Science and Engineering **856** (2020) 012005 doi:10.1088/1757-899X/856/1/012005

2016 *Curr. Top Med. Chem.* **16** 1951
[6] Bonnefont-Rousselot D 2016 *Nutrients* **8** 250
[7] Wright JS, Johnson ER, and DiLabio GA 2001 *J. Am. Chem. Soc* **123** 1173
[8] Tatefuji T, Yanagihara M, Fukushima S, and Hashimoto K 2014 *Food Chem Toxicol.* **67** 230
[9] Kato E, Tokunaga Y, and Sakan F 2009 *J. Agric. Food Chem.* **57** 2544
[10] Frisch MJ, Trucks GW, Schlegel HB, Scuseria GE, Robb MA, Cheeseman JR, Scalmani G, Barone V, Mennucci B, Petersson GA, Nakatsuji H, Caricato M, Li X, Hratchian HP, Izmaylov AF, Bloino J, Zheng G, Sonnenberg JL, Hada M, Ehara M, Toyota K, Fukuda R, Hasegawa J, Ishida M, Nakajima T, Honda Y, Kitao O, Nakai H, Vreven T, Montgomery JA Jr., Peralta JE, Ogliaro F, Bearpark M, Heyd JJ, Brothers E, Kudin KN, Staroverov VN, Kobayashi R, Normand J, Raghavachari K, Rendell A, Burant JC, Iyengar SS, Tomasi J, Cossi M, Rega N, Millam JM, Klene M, Knox JE, Cross JB, Bakken V, Adamo C, Jaramillo J, Gomperts R, Stratmann RE, Yazyev O, Austin AJ, Cammi R, Pomelli C, Ochterski JW, Martin RL, Morokuma K, Zakrzewski VG, Voth GA, Salvador P, Dannenberg JJ, Dapprich S, Daniels AD, Farkas Ö, Foresman JB, Ortiz J, Cioslowski and Fox DJ, Gaussian 09, Revision E.01, Gaussian, Inc., Wallingford CT, 2010.
[11] Nenadis N and Stavra K 2017 *J. Phys. Chem. A* **121** 2014
[12] Mikulski D, Gorniak R, and Molski M 2010 *Eur. J. Med.* **45** 1015
[13] Benayahoum A, Amira-Guebailia H, and Houache O 2014 *Comput. Theor. Chem.* **1037** 1

Search for the maximal threshold Co content for the oxidation behaviour point of view for 25wt.%Cr-containing TaC-strengthened cast superalloys designed for use at 1200°C on long time

P Berthod[1,2], J P Gomis[2]

[1,2]Université de Lorraine
[1]Institut Jean Lamour, Campus Artem, 2 allée André Guinier, 54000 Nancy, France
[2]Faculty of Sciences and Technologies, 54500 Vandoeuvre-lès-Nancy, France

patrice.berthod@univ-lorraine.fr

Abstract. The base of chromia-forming cast superalloys is the most often constituted of nickel and/or cobalt. Both elements may be simultaneously present, for instance Co for promoting good mechanical properties and Ni for stabilizing the austenitic matrix and favouring good ductility and machinability. In service, superalloys must also resist hot oxidation, property favoured by Ni but disadvantaged by Co. In (Ta,C)-rich cast superalloys exposed at high temperature, the intergranular strength may be enhanced by the very efficient TaC carbides while the higher the Co content the stronger the matrix creep resistance. In this work, six alloys designed to be chromia-forming and TaC-strengthened were exposed to laboratory air at 1200°C for 170 hours, to examine how the deterioration by oxidation and the evolution with time of the strengthening carbides in the bulk may be influenced by the Co/Ni ratio in the case of this important family of superalloys. The characterization of the formed oxide scales and of the morphologic evolution of the carbides allowed identifying a {0.6Co-0.4Ni}-proportion for the base elements. Such combination obviously allows keeping a chromia-forming behaviour and a good adherence of the protective oxide scale on the alloy, as well as a good resistance of the TaC against fragmentation.

1. Introduction

The superalloys based on nickel and/or cobalt represent the major family of metallic refractory alloys used on long times at high temperature in severe chemical and mechanical conditions [1]. Despite the more or less recent new ways of elaboration (additive manufacturing, single-crystalline solidification), fabrication by conventional casting still exists and keeps on offering high performance superalloys for geometrically complex components [2]. The high degree of interconnection of their matrixes and the equiaxed obtained structures allow high and isotropic creep strength, properties that remain interesting for many applications for which the high performance microstructures of γ/γ' single crystals cannot be successfully achieved [3]. Many turbine disks (and even some blades) for aeronautics or power generation as well as the most of the high speed rotating glass forming tools are still made of superalloys of this kind.

Being solid solution strengthened or reinforced by the intra-dendritic fine and dense precipitation of hard particles (carbides, intermetallic compounds) is compulsory for achieving high creep resistance of the matrix. For instance heavy atoms such as tungsten may improve this property in cast nickel-

based superalloys [4]. But tantalum, and also cobalt despite its lower atomic mass, may play such role. To complete the resistance against high temperature creep, the interdendritic spaces and grain boundaries of the equiaxed cast superalloys must be strengthened by hard and stable coarse particles interlocking neighbor dendrites. This can be successfully achieved by TaC carbides, but more in Co-rich superalloys [5,6] than in Ni-rich superalloys [7]. Co-rich nickel-based or cobalt-based superalloys obtained by conventional casting take benefit from matrix strength due to cobalt and efficient TaC-strengthening enhanced by the high presence of Co. On the other hand it is well known that the oxidation behaviour at high temperature of these alloys, necessarily based on chromium rather than on aluminium, is far from excellent but it can be improved by addition of Ni [8,9].
So, the main question for conventionally cast superalloys based on cobalt and/or nickel and the high temperature strength of which is based on solid solution hardening by Co and Ta dissolved atoms and on the interlocking by TaC carbides, is: what can be the Co/Ni ratio leading to the best compromise of high temperature mechanical and chemical behaviours? The purpose of the present study is to try responding this question for alloys containing 25wt.%Cr for favouring a chromia-forming oxidation behaviour and 0.4 wt.%C and 6 wt.%Ta for favouring the presence of TaC interdendritic carbides in optimized proportion.

2. Materials and methods

Six alloys were previously elaborated by high frequency induction melting for a thermodynamic study [10], with the form of 40g-weighing ingots. These alloys were designed to contain 25wt.%Cr, 0.4wt.%C and 6wt.%Ta. The remaining 98.6wt.% were composed of:

- only nickel for the “0CN5 Ta” alloy and only cobalt for the “5CN0 Ta” alloy
- n fifths of Co and (5-n) fifths of Ni for the “nCN(5-n) Ta” alloys named from “4CN1 Ta” to “1CN4 Ta”

The global chemical compositions and the as-cast microstructures of these alloys were controlled again, on parts taken in the same location in ingot as the samples for the oxidation tests of this study. This was carried out using Scanning Electron Microscopy (SEM) and Energy Dispersive Spectrometry (EDS).
Six other parts, each of about 10g with a planar face ground with #1200-grit SiC papers, were placed in a muffle furnace, each in a fairly large alumina shuttle. The applied heating was of 20°C/min up to 1200°C. The duration of the isothermal stage at 1200°C was 170 hours. It was finished by a cooling in the closed furnace after shut-off.
After return at room temperature, each oxidized sample was carefully handled and embedded in a metallographic cold resin mixture. For each oxidized sample, the parts of the external oxide scale which spalled off during the cooling were collected and subjected to X-ray diffraction (XRD) with a Philips X’Pert Pro diffractometer (Cu Kα, l = 1.541 Å) for their identification.
The embedded samples were cut into two parts. Each half was ground and polished until a mirror-like state. The obtained metallographic samples were examined by SEM in Back Scattered Electrons (BSE) mode using a JEOL JSM-6010LA electron microscope. Spot EDS analyses were carried out to identify the oxide natures in the parts of the external scale still attached to the alloy as well as the internal oxides. Spot analyses were also performed to value the chemical composition in the alloy close to the alloy/external scale interface, with special attention to the local contents in Cr and Ta.

3. Results & discussion

3.1. Microstructures

The chemical compositions of the alloys were found to be very close to the wished contents in Co, Cr, Ta and Ni. This verification of the C content was not possible, because of the too low molar mass of this element and of its targeted low content (0.4wt.%); however, by considering first that the applied elaboration protocol usually leads to a very good respect of the obtained C content by comparison to the wished one, and second that the obtained quantity of carbides is typical of this C content, it was assumed that all the six alloys really contain very close to 0.4wt.%. Concerning the initial microstructures of the alloys (figure 1), they are themselves totally similar to the ones of the alloys previously studied [10, 11]. To resume them, one can say that all the alloys are composed of a dendritic matrix and of carbides of one or two types, forming a eutectic compound with matrix in the interdendritic spaces. The three nickel-richest alloys (upper micrographs: "0CN5 Ta", "1CN4 Ta" and "2CN3 Ta") contain acicular-shaped eutectic chromium carbides (of the M_7C_3 type) and Chinese script eutectic tantalum carbides (TaC) while the three cobalt-richest alloys approximately contain only script TaC carbides.

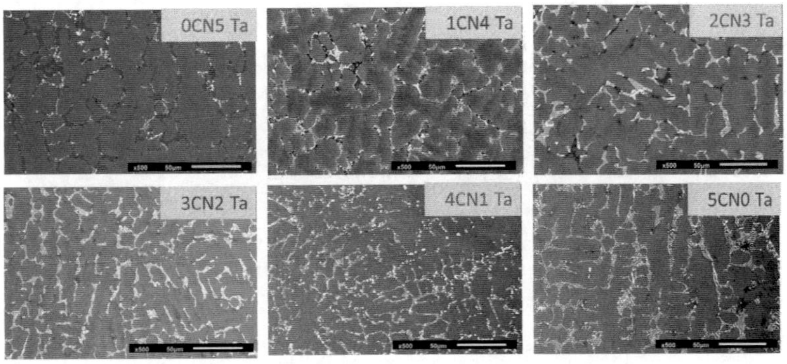

Figure 1. The initial microstructures of the alloys

Before examining the high temperature corrosion products one may have a look to the new bulk microstructures of the {1200°C, 170h}-exposed samples (figure 2). In the core of these samples, the microstructures were possibly changed, because of the high temperature stay but not altered by the surface oxidation reactions. The carbides morphologies were more or less modified. The chromium carbides significantly changed for becoming coarse globular carbides ("0CN5 Ta" and "1CN4 Ta" alloys) while they totally disappeared in the "2CN3 Ta" alloy which contain only TaC carbides with here and there curious shapes. The script-like TaC carbides initially present in the two Ni-richest alloys themselves turned into globules (fragmentation followed by coarsening). In contrast, the script-like tantalum carbides well resisted this morphologic evolution tendency, particularly the "'5CN0 Ta" alloy. Knowing the deleterious effect of the fragmentation of the carbides and of their coarsening as globules on the mechanical properties at high temperature, these observations suggest that the Ni-richest alloys may have lost the main part of their resistance against creep, in contrast with the Co-richest alloys the creep-resistance of which can be expected to stay at a high level.

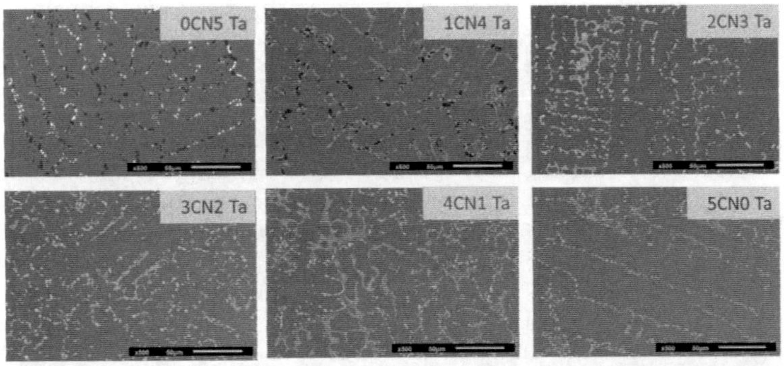

Figure 2. The bulk microstructures after the {1200°C, 170h}-stay

3.2. Oxides formed externally and internally

Concerning the high temperature corrosion products, the first important observation which can be made concern the more or less intense loss of the external oxide scale during the cooling. Obviously, the three Ni-richest alloys were particularly affected by this phenomenon since only some rare parts of the oxide scale still existing near the end of the isothermal stage can be observed on the metallographic samples. In contrast, the oxide scale seems having been more adherent to the surface of the Co-richest alloys. Thus, to totally specify the oxides formed during the 170 hours spent at 1200°C it was really necessary to analyse both the oxides remaining in the shuttle and the oxides still attached to the samples. The X-ray diffractograms performed on the mix of oxide parts collected in the ceramic shuttle after test allowed identifying the formed oxides belonging to the spalled-off parts of the oxide scales (figure 3: example of the "1CN4 Ta" alloy). This was completed by the cross-sectional characterization (BSE imaging and EDS analyses) on the embedded and polished samples (figure 4). The results of XRD and SEM/BSE identification of the oxides formed over the samples and in the sub-surfaces are summarized in table 1.

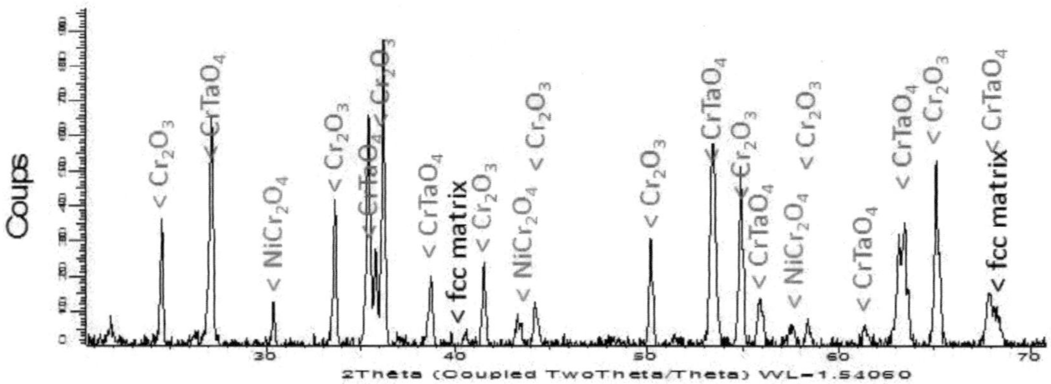

Figure 3. Diffractogram of the oxides detached from the 1CN4 Ta alloy

The diffractograms resulting of the XRD analyses of the six oxide powders collected in the six shuttles all show that chromia (Cr_2O_3) and the oxide of Cr and Ta ($CrTaO_4$) are present in all cases. The spinel oxides ($NiCr_2O_4$ for "0CN5 Ta" only, $CoCr_2O_4$ for the five other alloys) also formed systematically. One can also note the presence of NiO which formed probably as an outer part of the external scale, as this is usual for Ni-based alloys. The oxides still present on the samples after return to ambient temperature are more complex. They mainly concern the three Co-richest alloys. The best illustration is given by the cobalt-based "5CN0 Ta" alloy (micrograph in figure 4 and oxide description in table 1). In addition it is obvious that the higher the Co content in these Co-richest alloys, the higher the scale thickness and the more complex the composition of the scale.

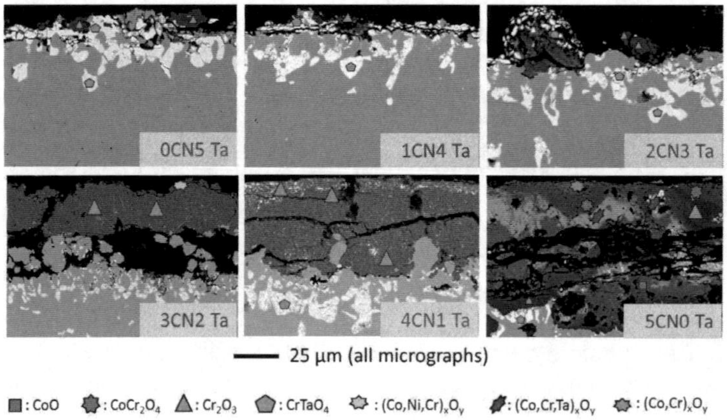

Figure 4. Characterization of the oxidized states of the alloys' surfaces

Table 1. Synthesis of the oxides' identifications

	X-Ray diffraction	SEM/EDS
0CN5 Ta	NiO, $NiCr_2O_4$	/
1CN4 Ta	$NiCr_2O_4$	/
2CN3 Ta	$CoCr_2O_4$	/
3CN2 Ta	$CoCr_2O_4$	$(Co,Ni,Cr)_xO_y$, $CoCr_2O_4$
4CN1 Ta	$CoCr_2O_4$	(Co,Ni)O, $CoCr_2O_4$, $(Co,Ni,Cr)_xO_y$
5CN0 Ta	$CoCr_2O_4$	CoO, $CoCr_2O_4$, $(Co,Cr)_xO_y$, $(Co,Cr,Ta)_xO_y$
+ in all cases:	Cr_2O_3 and $CrTaO_4$	Cr_2O_3 and $CrTaO_4$

The chemical composition of the outermost parts of the alloys (close to the alloy/oxide scale interface) was measured by EDS spot analysis. The results concerning Cr and Ta are both plotted in figure 5. It obviously appears that the higher the Co content in the alloy the lower the Cr content in extreme surface. The decrease in Cr is furthermore more marked when the alloy becomes Co-based instead Ni-based (the Cr content becomes lower than 15wt.% for the "4CN1 Ta" alloy). For tantalum, the Ta decrease is the fastest when Co starts to replace Ni in the Ni-richest alloy.
These results show that these are now the Co-richest alloys which present the worst behaviours. Their oxide scales are composed of multiple oxides, much less protective than the chromia layer which efficiently prevented fast oxidation of the Ni-richest alloys, as suggested by the Cr and Ta contents in extreme surface remained rather high. The loss of the chromia-forming behaviour of the Co-richest alloys, resulting in a general oxidation of all the base elements into single or mixed oxides, is in good agreement with the Cr contents on extreme surface become particularly low.

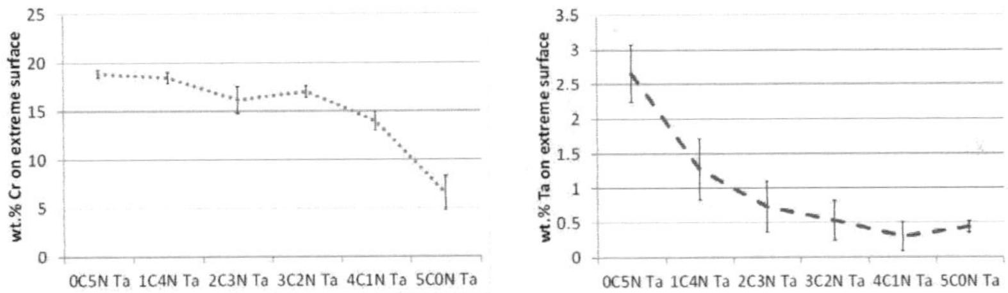

Figure 5. The Cr (left) and Ta (right) contents in extreme surface

4. Conclusion

Globally, one can consider that the TaC-reinforced {Co, Ni}-based alloys behaved rather well during the applied oxidation test, by considering the particularly high level of temperature and long exposure. The currently best superalloys, the γ/γ' Ni-based single-crystals, are much better in term of oxidation resistance but lose their excellent creep resistance as soon as the service temperature becomes higher than 1100°C. Some of the present alloys – the Co-richest ones – are obviously able to keep almost intact their script-like TaC carbides at 1200°C, even until 170 hours of isothermal exposure. The problem is that their oxidation resistance is not good enough. Fortunately, adding Ni in sufficient quantity (2/3 of Co replaced by Ni: "3CN2 Ta" alloy) may lead to a not perfect but much better oxidation resistance without significantly deteriorating the population of the strengthening TaC carbides. The investigation of the performances of this alloy merits to be deepened, notably by real creep tests at this temperature of 1200°C.
To finish, one can remark that the influence of the Ni/Co ratio, notably the beneficial effect of Ni on the high temperature oxidation resistance observed here, is in good agreement with observations made in previous works. What needs to be underlined here is that the present results especially concern new alloys which are especially rich in Ta (6wt.%, much higher than many superalloys) and designed to

contain exclusively TaC carbides by the choice of the same molar contents for Ta and C (chemical design which appears here to be successful only for Ni contents not too high).

Acknowledgments

The authors wish thanking G Medjahdi, P J Panteix and L Aranda for their technical help.

References

[1] Sims C T and Hagel W C 1988 The Superalloys (New York: John Wiley & Sons)
[2] Donachie M J and Donachie S J 2002 Superalloys: A Technical Guide (Materials Park: ASM International)
[3] Durand-Charre M 1997 The Microstructures of Superalloys (Boca Raton: CRC Press)
[4] Zhou T, Ding H, Ma X, Feng W, Zhao H, Meng Y, Zhang H and Lv Y 2018 *Mater. Sci. Eng.* A **725** 299
[5] Michon S, Aranda L, Berthod P and Steinmetz P 2004 *Rev. Mater./C.I.T.* **101** 651
[6] Montazeri M, Ghaini F M and Farnia A 2018 *Int. J. Mater. Res.* **102** 1446
[7] Qin X Z, Guo J T, Yuan C, Hou J S and Ye H Q 2008 *Mater. Letters* **62** 2275
[8] Kofstad P 1988 High Temperature Corrosion (London: Elsevier applied science)
[9] Young D J 2008 High Temperature Oxidation and Corrosion (Amsterdam: Elsevier corrosion series)
[10] Berthod P and Himeur Z 2018 *Calphad* **60** 156
[11] Berthod P and Himeur Z 2018 *Oxid. Met.* **90** 187

IOP Conf. Series: Materials Science and Engineering **856** (2020) 012007 doi:10.1088/1757-899X/856/1/012007

Carbon nanotube reinforced aluminium composites fabricated by powder metallurgical and hot-rolling techniques

Hideaki Tsukamoto

Department of Mechanical Engineering, Hosei University,
3-7-2 Kajino-cho, Koganei, Tokyo 184-8584, Japan

htsukamoto@hosei.ac.jp

Abstract. This research aims to fabricate light weight and high strength materials such as multi-walled carbon nanotubes (CNT) reinforced aluminium composites at relatively low cost. CNT/ Al composites have been fabricated in three steps: (1) preparation of CNT dispersion, (2) preparation of CNT/ Al based composite material by powder metallurgical slurry method, and (3) strengthening and refining of CNT/ Al based composites by rolling processes. After fabrication, the composites have been characterized on dispersibility of CNT, Vickers hardness, three-point bending and tensile properties.

1. Introduction

In recent years, in order to cope with environmental problem issues such as CO_2 emission reduction, replacing various conventional structural materials like heavy metals consisting major parts in automobiles such as iron materials with light metals such as aluminum (Al), magnesium (Mg) and titanium (Ti) have been desirable. However, the weight reduction of structural materials by light metals often leads to decrease in strength, which causes problems threatening securing safety and sustainability. Attempts to develop carbon nanotube (CNT)/ Al-based composites with enhanced strength have been highly attractive. However, at present there are very few practical examples due to difficulty to obtain uniform dispersion and wetting of CNTs with the matrix [1,2].

Fillers usually need some surface treatment to improve dispersion and wetting, but it is difficult to uniformly treat nanomaterials such as CNTs. Further, it has been reported that there is a possibility that the properties of the CNT may be impaired by the surface treatments. It was proposed that in order not to reduce the properties of CNT, dimethylacetamide was used as a solvent, and potassium carbonate which is an inorganic salt as a dispersing agent. Dispersion of CNT in the solvent was performed by sonication [3].

In this study, CNTs with high modulus and strength have been uniformly dispersed in Al matrix phase, and high-performance light-weight CNT / Al-based composites have been fabricated by combining powder metallurgy and hot rolling techniques. In addition, the relation between microstructures and mechanical properties have been investigated based on examination of electron microscope images, micro Vickers hardness, three-point bending and tensile test results.

2. Experiments

2.1 Fabrication

Al powder with a particle size of 30 (μm) and multi-walled carbon nanotubes (MWNT) with a diameter of 10-15 (nm) and an aspect ratio of 1000 were mixed. Potassium carbonate was used as dispersant, and dimethylacetamide was used as solvent. The CNT powder and potassium carbonate were put in dimethyl acetamide and sonicated with an ultrasonic equipment. The mixture was filtered using filter paper, and the CNT powder was taken out. A mixed powder was prepared from the CNT and Al powders, poured into a mold and dried using a heat gun. After that, pressure molding was performed at 150 kPa using a press machine. Then, it was sintered at 500 ° C for 2 hours in an electric furnace.

2.2 Hot rolling

The test pieces prepared were heated in an electric furnace at 400 ° C for 15 minutes, and repeated hot rolling was performed with a reduction ratio of 3 % for a rolling process at the beginning. After ten time rolling, the reduction ratio was changed to 4 to 5 % for a rolling process. Then, the rolling was repeatedly conducted to total reduction ratio of 30 %. Fig.1 shows the repeated processes of hot rolling.

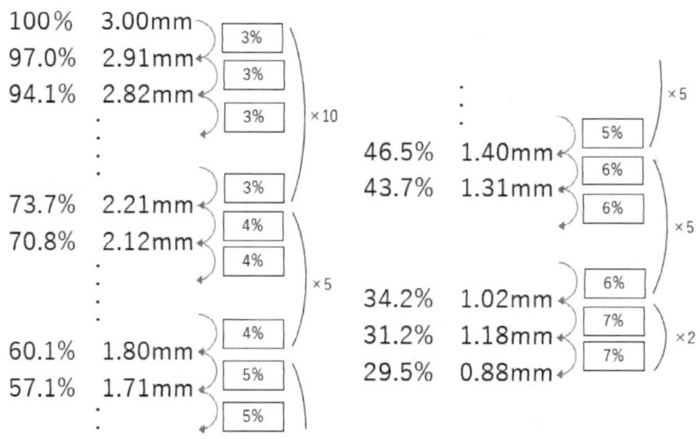

Figure 1. Repeated hot rolling processes

2.3 Characterization

Evaluation of dispersibility of CNT in the composites, Vickers hardness, three-point bending and tensile tests were carried out. Cross-sectional observations in CNT/ Al composites was performed using a scanning electron microscope (SEM) and energy dispersive X-ray spectrometry (EDX).

For Vickers hardness test, the cross sections of the test pieces were polished, and the Vickers hardness test (on five points) was performed. For three-point bending tests, the test pieces were prepared by cutting out edge sides from as-rolled samples. Displacement rate was set at 1.0 mm/ min in the bending tests. The support span is 25mm. For tensile tests, Instron-type tensile equipment was used, and the strain rate was set at $5.6x10^{-6}$/ sec. Elastic modulus, tensile strength and maximum tensile strain were estimated. Anisotropy was also examined.

IOP Conf. Series: Materials Science and Engineering **856** (2020) 012007 doi:10.1088/1757-899X/856/1/012007

3 Results and Discussion

The SEM backscattered electron image and EDX image of the cross section of CNT/ Al composites are shown in Fig.2.

(a) (b)

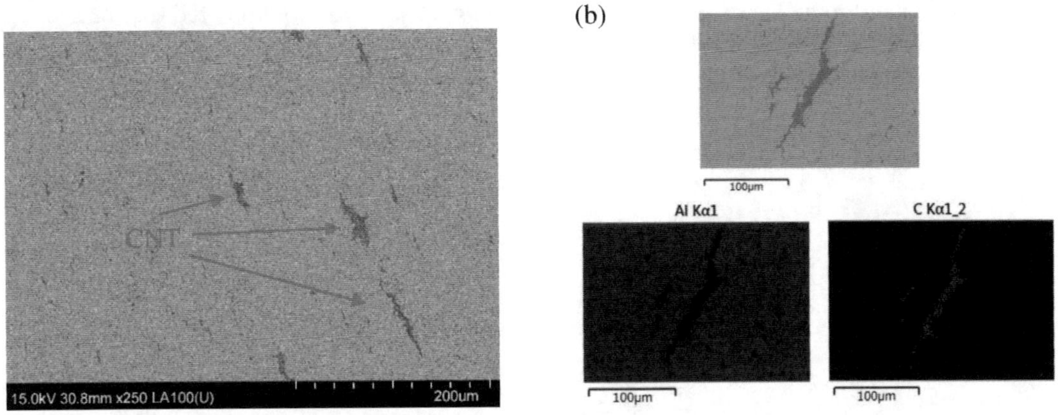

Figure 2. Cross-sectional Observation of CNT/Al Composites (Ultrasonic Treatment for 1 hour, Carbon Nanotube Content of 1.0 mass .%) (a) SEM image, (b) EDX image

CNTs are seen to be relatively uniformly dispersed. Some aggregation of CNT can be also observed. The photos of the test pieces before and after rolling are shown in Fig. 3.

(a) (b)

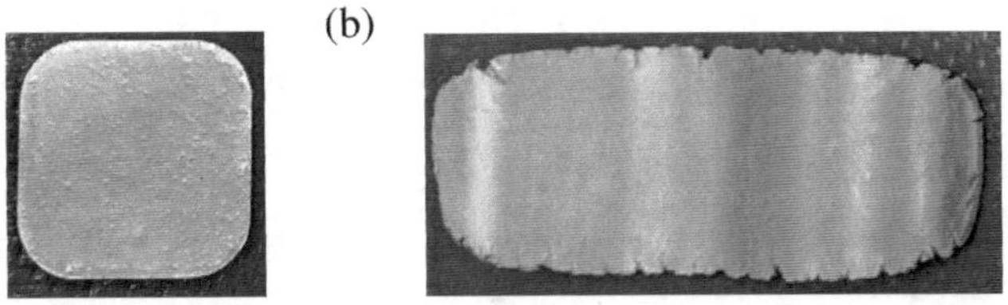

Figure 3. Sample apparatus (a) before and (b) after rolling

Fig.4 shows the relationship between bending stress and strain (extracted from three-point bending test results) for sonication time of 0h, 3h, and 4h (hour). The content of CNT was 0.5 mass %.

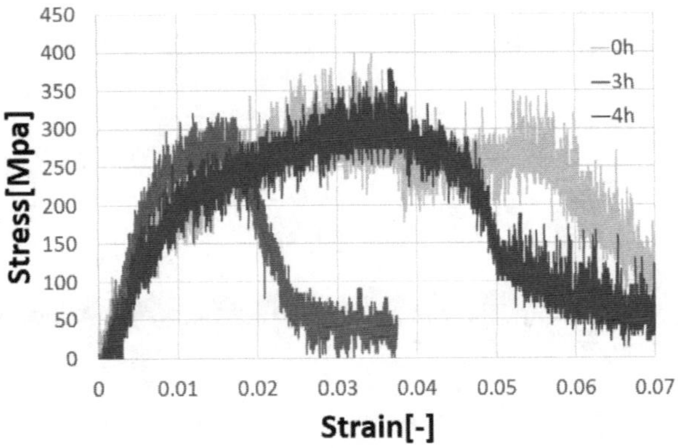

Figure 4. Stress-strain diagram for CNT (0.5 mass %) / Al composites from three-point bending tests

Fig.5 shows the relationship between sonication time and flexural modulus. The CNT content was 0.5 mass %. The flexural modulus obtained from the bending tests are lower than the values in literature. It can be confirmed that the flexural modulus increases as sonication time increases to sonication time of 3h. In case of 4h, the flexural modulus sharply drops to the value which is the almost same as that for 0h. It can also be seen that ductility is significantly impaired for 3h. The reason for the low value of flexural modulus at 0h is due to the least dispersion of CNTs forming CNT aggregation and lumps. Some voids exist, which leads to decrease in flexural modulus.

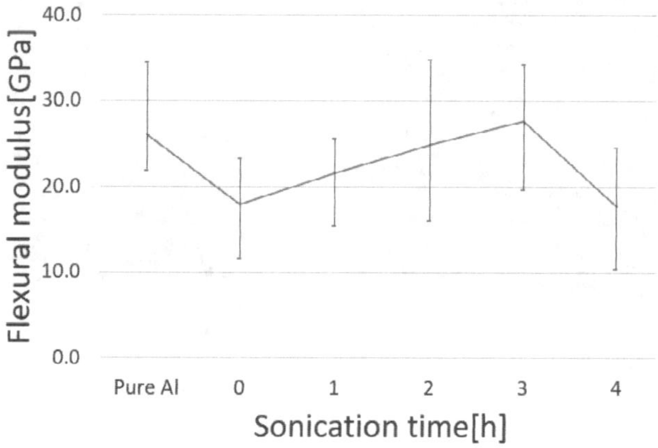

Figure 5. Relationship between flexural modulus and sonication time.

Next, we move to the results of tensile test. Fig. 6 shows stress-strain curves for pure Al and CNT/ Al composites with sonication time of 0h, 1h, 2h, 3h and 4h. The CNT content was 0.5 mass%.

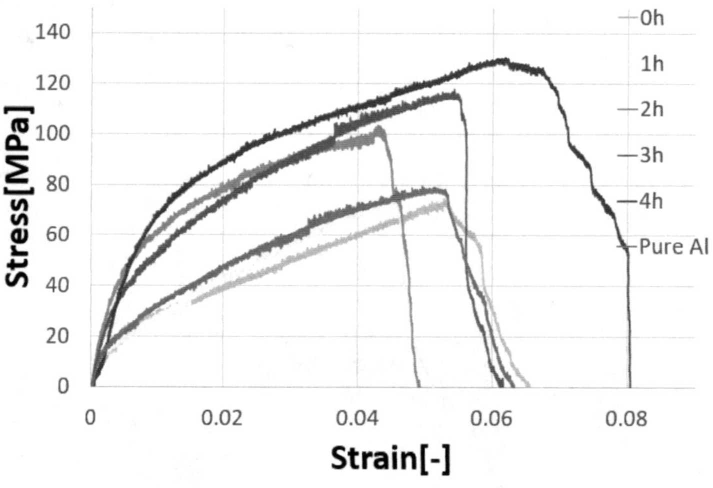

Figure 6. Stress-strain diagram for CNT (0.5 mass %)/ Al composites.

The Young's modulus and tensile strength as a function of sonication time are shown in Fig.7 (a) and (b). The CNT content of the composites was 0.5 mass%.

IOP Conf. Series: Materials Science and Engineering **856** (2020) 012007 doi:10.1088/1757-899X/856/1/012007

(a)
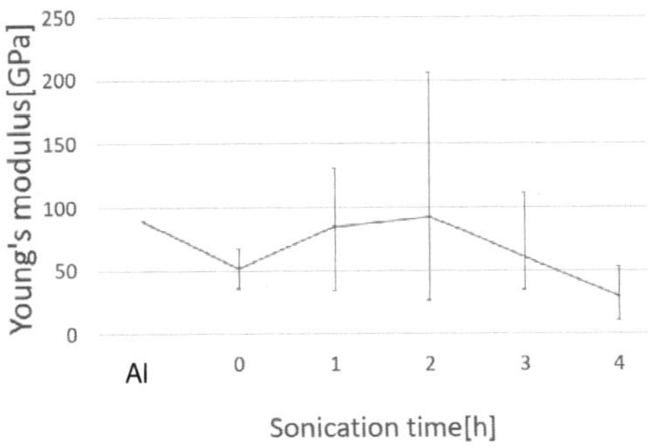

(b)
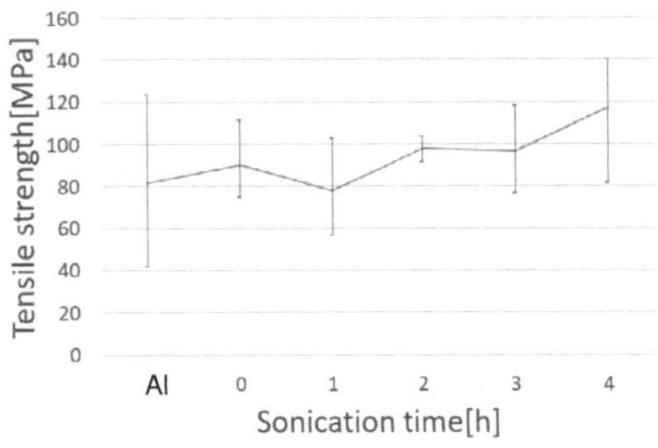

Figure 7. Relationship between (a) tensile strength, (b) Young's modulus and sonication time.

Elastic modulus of pure Al is seen around 90 GPa, and average values of elastic modulus of the composites are between 30 and 92 GPa. Maximum value is achieved for the composites with sonication time of 2h, and minimum is for the composites with sonication time of 4h. There can be highly scattering data in elastic modulus of the composites with sonication time of 2h. For tensile strength, with sonication time increasing, tensile strength increases. The composites with sonication time of 4h shows highest value of the tensile strength. It is also seen in Fig.7 (b) that the composites with sonication time of 4h achieved highest tensile strength and deformation ability.
Fig.8 shows Vickers hardness test results. The CNT content was 0.5 mass%.

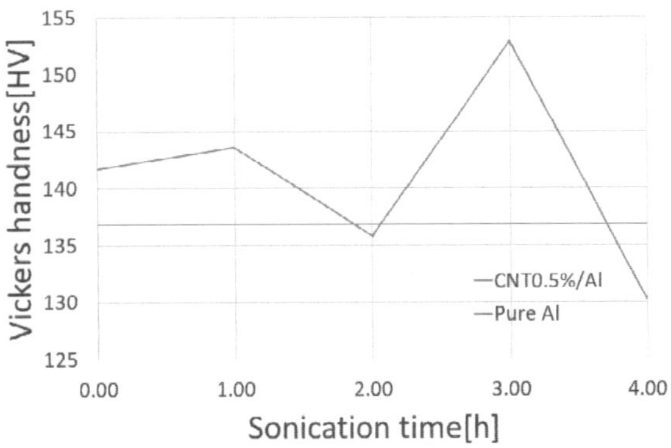

Figure 8. Relationship between Vickers hardness and sonication time

The literature value of Vickers hardness of pure Al (1000 series) is around 30 (HV). In this study, the rolled Al shows more than 4 times of the hardness, because work hardening occurs as plastic deformation undergoes by rolling. It is also seen that the influence of sonication time is not so large, and CNT does not have large effect on Vickers hardness.

4 Conclusions

1. CNT/ Al based composites were fabricated by a technique combining powder metallurgy and repeated hot rolling.
2. The tensile strength of the composites was enhanced by dispersing of CNTs in Al matrix, which was achieved with otassium carbonate in dimethyl acetamide by sonication in an ultrasonic bath.
3. It was found that the sonication time had a great influence on dispersing state of CNT.
4. Vickers hardness of the composites was not highly influenced by sonication time and CNT content.
5. Three-point bending test results demonstrated that flexural modulus increased with sonication time increasing up to 3h.
6. Tensile test results demonstrated that sonication time did not highly affect elastic modulus, while tensile strength increased as sonication time increased.

References

[1] Yoshitoshi Nakayama, Functionalization and Composite Technology of Carbon Nanotubes, **p. 214-216**, 2006.
[2] Hashim Hanizam, Mohd Shukor Sallehc, Mohd Zaidi Omar, Abu Bakar Sulong, J.Mater.Res.Technol., 8 (2), **p.2223–2231**, 2019.
[3] Chenglin Yi, Xiaoming Chen, Feilin Gou, Christopher M. Dmuchowski, Anju Sharma, Cheol Park, Changhong Ke, Carbon 125, **p.93-102**, 2017.

IOP Conf. Series: Materials Science and Engineering **856** (2020) 012008 doi:10.1088/1757-899X/856/1/012008

Micro/nanostructured MTMOs as Electrocatalysts for Oxygen Evolution Reaction in Alkaline Medium

Charles Lois Flores and Mary Donnabelle Balela

Sustainable Electronic Materials Group, Department of Mining, Metallurgical, and Materials Engineering, University of the Philippines, Diliman, Quezon City, Philippines 1101

ciflores@up.edu.ph, mlbalela1@up.edu.ph

Abstract. Different nanostructured mixed transition-metal oxides were synthesized using a facile and controllable hydrothermal method. Incorporation of Cu^{2+} and Ni^{2+} in Co_3O_4 structure led to an interesting morphological transformation, from quasi-spherical nanoparticles to microspherical urchin-like aggregates of radially assembled nanoneedles. These oxides have notable low crystallite size, narrow size distribution, and high degree of crystallinity. In general, partial cobalt substitution resulted in an improved electrocatalytic activity for oxygen evolution reaction (OER), with $NiCo_2O_4$ demonstrating superior performance among the catalysts prepared. It recorded a low overpotential of 374 mV and achieved high stability, making it competitive with state-of-the-art RuO_2 and IrO_2 (387 mV). This material is of great potential for sustainable energy applications.

1. Introduction

Noble metal oxides, such as ruthenium and iridium oxide (IrO_2 and RuO_2), remain to be the key electrocatalysts for oxygen evolution reaction (OER) owing to their remarkable electrochemical properties [1]. However, these materials are known to be expensive, nonnaturally occuring, and unstable in alkaline conditions. These impede their use for large scale commercial applications. Such limitations have attracted interest in cheaper and abundant transition metal oxides as new electrocatalysts. In particular, cobaltosic oxide (Co_3O_4) has attracted huge interest predominantly due to its due to its high redox activity and enhanced stability [2]. The excellent electrochemical stability is attributed to the ease in which cobalt (Co) ion can shift between oxidation states. Furthermore, it has been shown that partial Co substitution by other transition metals (M = Mn, Ni, Cu, Fe, Zn) producing MCo_2O_4, reduced the cost and toxicity while enhancing the electrochemical performance, through the synergistic effect of individual components [3]. Therefore, various Co-based mixed transition metal oxides (MTMOs) have been synthesized through different techniques such as solvothermal method, thermal decomposition, solution combustion, coprecipitation, and sol-gel process [4-7]. However, these synthesis routes typically require expensive equipment and aggressive experimental conditions. Most of the fabricated Co-based MTMOs of diverse morphologies were also studied for supercapacitor applications only. There also have been no reports about the effects of incorporating transition metal ions in Co_3O_4 on the resulting morphology of the MTMOs.

In this work, different cobalt-based MTMO nanostructures were fabricated using a facile two-step ethanol-assisted hydrothermal method. Effects on the morphology, structure, electrochemical performance, and stability of the electrocatalysts upon partial Co substitution of Ni^{2+} and Cu^{2+} in

Co_3O_4 were investigated. Further, the electrolyte concentration was varied to determine its effect on the electrocatalytic behavior of the prepared electrodes.

2. Experimental process

A mixture of 2.0 g urea (NH_2CONH_2, Techno Pharmchem Haryana) and 0.5 g cetyltrimethylammonium bromide (99%, CTAB, Loba Chemie Pvt. Ltd.) in 20 mL deionized water and 10 mL ethanol (CH_3CH_2OH, Univar) was stirred for 10 min. Four millimoles of cobalt (II) nitrate hexahydrate ($Co(NO_3)_2 \cdot 6H_2O$, Sigma-Aldrich) was added to produce Co_3O_4. This was further mixed with 2 mmol of either copper (II) nitrate hemi(pentahydrate) ($Cu(NO_3)_2 \cdot 2.5H_2O$, Sigma-Aldrich) and nickel (II) nitrate hexahydrate ($Ni(NO_3)_2 \cdot 6H_2O$, RTC Laboratory) to produce different cobalt-based MTMOs. The resulting solution was stirred for 30 min and then transferred to a Teflon-lined stainless steel autoclave reactor. Hydrothermal reaction was carried out in a vacuum oven at 100 °C for 10 h. Thereafter, the precipitate was collected and washed several times with deionized water and ethanol (CH_3COOH, Univar) through centrifugation, and then dried in an oven at 60 °C for 6 h. The as-synthesized precursors were calcined in air at 400 °C for 6 h with a heating rate of 1 °C/min.

The size and morphology of the synthesized oxides were observed using field emission scanning electron microscope (FESEM, Hitachi SU-8230). The length and diameter of the nanostructures were measured using Image Processing Analysis in Java software (ImageJ). The crystal structure was analyzed through X-ray diffraction (XRD, Shimadzu XRD-7000) using Cu K_α (λ=1.5406 Å) while the specific surface area was obtained through Brunauer-Emmett-Teller analysis (BET, Quantachrome Instruments).

The electrochemical performance of the samples was examined using a conventional three-electrode system with 1 M potassium hydroxide (KOH, Sigma-Aldrich) electrolyte. The concentration of KOH was varied from 0.01 to 1.0 M to determine its effect on the electrochemical performance of the catalysts. Ag/AgCl (Autolab 6.0726.107) and platinum (Pt) wire were employed as reference and counter electrodes. A piece of carbon paper drop-cast with the catalyst ink [90% catalyst sample, 10% polytetrafluoroethylene (PTFE, Sigma-Aldrich) binder, and a few drops of ethanol] was used as the working electrode. The mass loading of the prepared catalysts was about 1 mg/cm^2. Linear sweep voltammetry (LSV) was performed in the potential range of 0 to 1 V at a scan rate of 1 mV/s to avoid capacitive currents during the measurement. The onset and overpotential were calculated at 1 and 10 mA/cm^2 current density (j), respectively. Furthermore, to assess the stability of the catalyst in OER, cyclic voltammetry (CV) was carried out at a scan rate of 100 mV/s in the range of 0.4 to 0.8 V for 1000 cycles. All potential values were reported against the reversible hydrogen electrode (RHE).

3. Results and discussion

3.1. Morphological analysis

Displayed in Fig. 1 are the FESEM images of the synthesized (a) Co_3O_4, (b) $NiCo_2O_4$, and (c) $CuCo_2O_4$ nanostructures. Quasi-spherical nanoparticles with a mean diameter of about 63.03 nm were observed in pure (undoped) Co_3O_4 as seen in Fig. 1a-b. Partial substitution of transition metal ions resulted to a dramatic change in the morphology of the precipitates. Upon addition of Ni^{2+}, microspheres having a mean diameter of about 3.21 μm were formed. Higher magnification images revealed that the microspheres are actually composed of radially assembled nanoneedles with a mean cross-sectional diameter of 20.62 nm. Particles are also observed to lay on top of the urchin-like structures as seen in Fig. 1d. On the other hand, similar urchin-like precipitates (mean diameter of about 5.75 μm) were obtained when Cu^{2+} was added instead of Ni^{2+}. However, the nanoneedles appeared finer with a cross-sectional diameter of about 25.36 nm. This drastic evolution in morphology upon incorporation of transition metal ions could be attributed to a two-stage process. First, reaction among the precursors yielded nanoparticles that have successively agglomerated to produce large microspheres through oriented attachment [8-9]. It is known that minute nanoparticles have a strong tendency to agglomerate, to minimize their surface energies [8]. Subsequently, incorporation of transition metal ions could provide favorable sites (low interfacial energy) for the absorption of the reactant molecules and direct the anisotropic growth [10,11] which possibly caused the formation of nanourchins. Furthermore, the formation of micro/nanosized structure could be attributed to the combined effect of ethanol and CTAB during synthesis. Ethanol, when mixed with water, produces an organic solution of low permittivity [12]. This allows the product to be liberated from foreign anions or other ionic species, thus, providing excellent chemical homogeneity. Consequently, nucleation and growth rate are controlled, leading to small particle size. On the other hand, CTAB reduces the surface energy that hinders the growth of the particles [12,13].

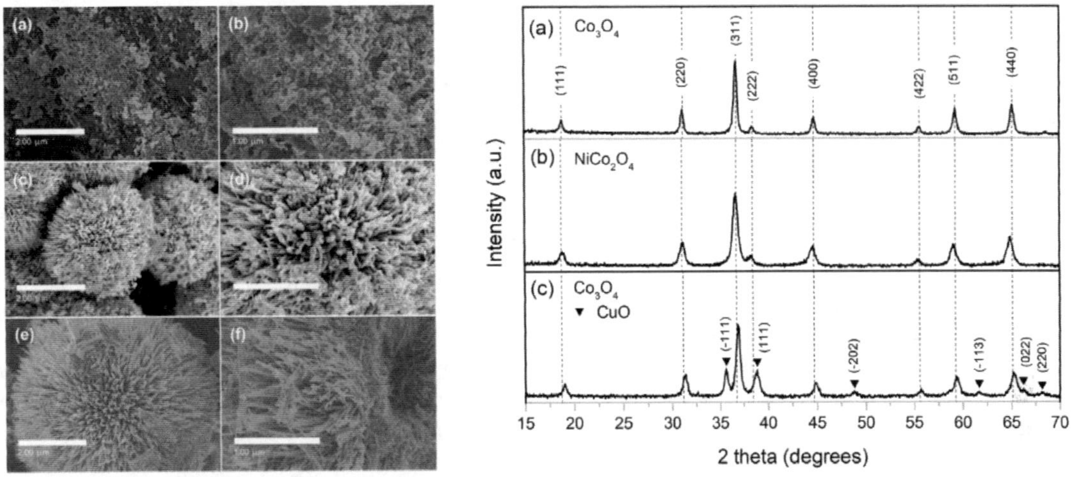

Figure 1. Low and high magnification FESEM images of (a-b) Co_3O_4, (c-d) $NiCo_2O_4$, and (e-f) $CuCo_2O_4$ after calcination at 400 °C (left) and the XRD patterns (right).

3.2. Structural analysis

The XRD patterns of the synthesized oxides after calcination at 400 °C are also presented in figure 1. Peaks at 2θ = 18.80, 31.15, 36.71, 38.38, 44.62, 55.52, 59.20, and 65.10° are indexed to 111, 220, 311, 222, 400, 422, 511, and 440 planes of Co_3O_4 (JCPDS Card No. 001-6025), respectively, confirming the formation of pure Co_3O_4. Interestingly, these peaks are also observed in the diffraction patterns of $NiCo_2O_4$ (JCPDS Card No. 001-6025) and $CuCo_2O_4$ (JCPDS No. 78–2177). No other peaks, except $NiCo_2O_4$, were observed in the sample with Ni, which indicates high purity. After doping with Ni^{2+}, the lattice parameter increased from 8.1001±0.0002 to 8.1168 ± 0.0002 relative to pure Co_3O_4 as shown in Table 1. This suggests the successful incorporation of Ni^{2+} in Co_3O_4 without altering the spinel structure. In the case of the sample with Cu^{2+}, a slight increase in lattice parameter was also observed. Peaks at 2θ = 35.61, 38.83, 48.85, 61.65, 66.62, and 68.20° were indexed to the -111, 111, -202, -113, 022, and 220 planes, respectively, of the monoclinic CuO (JCPDS No. 001-6025). This implies formation of a secondary phase possibly due to the oversaturation of the solid solution of Co and Cu ions, hence producing a $CuCo_2O_4$/CuO composite [9,13].

Additionally, it could be observed that the diffraction peaks become relatively wider after the incorporation of metal ions. This could be ascribed to the lower degree of crystallinity or the small crystallite size of the oxides [13], which ranges from about 12.25 to 20.43 nm. Consequently, higher BET specific surface area was recorded with the sample having smaller crystallite size. As shown in Table 1, the 30.83 m^2/g specific surface area of Co_3O_4 becomes 31.72 and 49.04 m^2/g, a rise of about 2.8 and 37.1 % upon introducing Cu^{2+} and Ni^{2+}, respectively.

Table 1. Crystallite size, lattice parameter, and BET specific surface area of the synthesized oxides.

Sample	Crystallite size (*nm*)	Lattice parameter (*Å*)	Specific surface area (m^2/g)
Co_3O_4	20.43	8.1001 ± 0.0002	30.83
$CuCo_2O_4$/CuO	15.17	8.1066 ± 0.0058	31.72
$NiCo_2O_4$	12.25	8.1168 ± 0.0002	49.04

IOP Conf. Series: Materials Science and Engineering **856** (2020) 012008 doi:10.1088/1757-899X/856/1/012008

3.3. Electrochemical performance

Shown in figure 2a are the LSV polarization curves of the electrocatalysts for OER. Based on the results summarized in Table 2, Co_3O_4 obtained an onset potential of 1.57 V vs RHE while $CuCo_2O_4/CuO$ and $NiCo_2O_4$ both recorded about 1.52 V vs RHE. Overpotential of the catalysts obeys a similar trend, with Co_3O_4, $CuCo_2O_4/CuO$, and $NiCo_2O_4$ registering about 468, 386, and 374 mV, respectively. These partially substituted transition metal oxides perform better than the commercially available RuO_2 (387 mV) and IrO_2 (387 mV) [14] in terms of overpotential. Furthermore, the current density attained at 1.75 V follows the order: $NiCo_2O_4$ (38.12 mA/cm^2) > $CuCo_2O_4$ (37.67 mA/cm^2) > Co_3O_4 (15.74 mA/cm^2). Same trend was observed for mass activity, with values ranging from about 7.01 to 17.84 A/g. This enhancement in electrocatalytic activity upon incorporation of other transition metal ions could be ascribed to the altered Co distribution and introduced vacancies [1] in the Co_3O_4 structure. Consequently, this leads to an increase in conductivity. In particular, $NiCo_2O_4$ demonstrated the most excellent electrocatalytic behavior possibly because of its very small crystallite size and high BET specific surface area. Aside from that, its porous nanourchin structure probably provides huge channels that allow effective penetration of the electrolyte, to facilitate the transport electrons during OER. On the other hand, the presence of CuO in $CuCo_2O_4/CuO$ composite could have a detrimental effect in its electrocatalytic property since CuO is highly resistive [9,13].

Further evaluation of the electrocatalytic behavior of each catalyst was conducted by estimating the Tafel slope from the corresponding LSV polarization curves. As presented in figure 2b, Co_3O_4, $CuCo_2O_4/CuO$, and $NiCo_2O_4$ achieved 95.1, 85.0, and 68.4 mV/decade, respectively. These values are relatively low for Co-based systems [9]. This finding supports that $NiCo_2O_4$, having the best OER kinetics, is the most efficient among the prepared catalysts. Its Tafel slope of about 68.4 is close to 60 mV/decade. This could signify that the rate determining step for water oxidation entails one electron transfer [1,9]. Moreover, this suggests that the most probable functional species in the electrocatalysis of OER is Co^{3+} [14]. Co^{3+} ions are situated in the more exposed octahedral sites of the spinel structure, hence facilitating their oxidation to Co^{4+}. In addition, presence of such high valent Co cations with a greater number of unpaired *d* electrons, could lower the activation energy for the chemisorption of gas to the surface of the catalyst. Therefore, both the formation of hydroperoxy species (OOH) and OER are hastened [15], as suggested in equation (3) and (4). The mechanism for OER [15,16] is shown below:

$$M + OH^- \rightarrow M - OH + e^- \quad (1)$$

$$M - OH + OH^- \rightarrow M - O + H_2O + e^- \quad (2)$$

$$M - O + OH^- \rightarrow M - OOH + e^- \quad (3)$$

$$M - OOH + OH^- \rightarrow M - O_2 + H_2O + e^- \quad (4)$$

where *M* is the active species. Meanwhile, based on this series of reactions, decreasing the amount of OH^- would inhibit the formation of O_2. This is clearly shown in figure 2d, wherein a dramatic reduction of current density (38.12 to 0.03 mA/cm^2) and an apparent shift to higher onset potentials were observed after lowering the KOH concentration from 1 to 0.01 M. On the other hand, stability is an equally important measure of a good electrocatalyst, especially for upscale applications in energy storage and conversion systems. As illustrated in figure 2c, all the catalysts showed good stability even after 1000 cycles, recording a quite minimal 4.13 to 4.88 % drop in current density at 1.75 mV.

Table 2. Summary of the electrocatalytic activity of the MTMOs in 1 M KOH.

Sample	Overpotential (*mV*)	Tafel slope (*mV/dec*)	*j* at 1.75 V (*mA/cm^2*)	Mass activity (*A/g*)
Co_3O_4	468	95.1	15.74	7.01
$CuCo_2O_4/CuO$	386	85.0	37.13	11.12
$NiCo_2O_4$	374	68.4	38.12	17.84

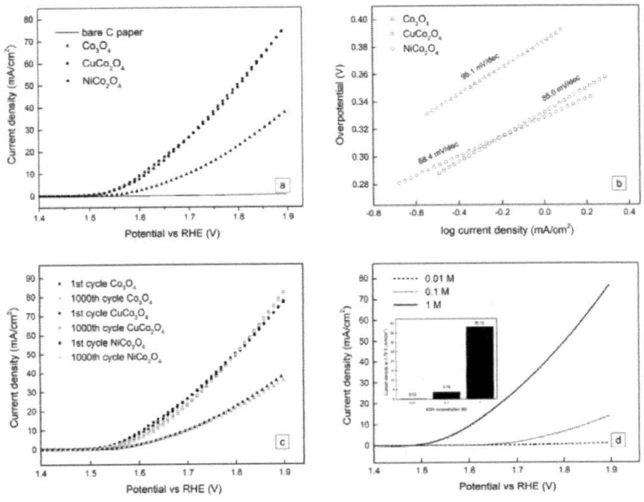

Figure 2. (a) LSV curves of the electrocatalysts in 1 M KOH, (b) corresponding Tafel plots, (c) LSV curves of the electrocatalysts before and after 1000 cycles in 1 M KOH, and (d) LSV curves of $NiCo_2O_4$ in 0.01 to 1 M KOH.

4. Conclusion

Different nanostructured cobalt-based MTMOs were fabricated using a facile two-step hydrothermal method. Initially, Co_3O_4 quasi-spherical nanoparticles were formed. Addition of Cu^{2+} and Ni^{2+} resulted to the formation of microspherical urchin-like aggregates of radially assembled nanoneedles. This dramatic morphological transformation could be accounted for the incorporation of metal ions in the Co_3O_4 structure. Remarkably, the nanostructures have regular shapes, micro/nanosized crystals, and high degree of crystallinity, which are possibly imparted by the introduction of ethanol and CTAB during synthesis. In general, partial cobalt substitution with other transition metal ions led to an enhanced electrochemical performance, with $NiCo_2O_4$ exhibiting superior activity. This material is of great potential for energy storage.

Acknowledgment

This work is supported by CHED-Newton Agham Program under the project entitled "Affordable Electrolyzer Technology based on Transition Metal Catalysts for Energy Storage Applications."

References

[1] Osgood H, Devaguptapu S V, Xu H, Cho J, and Wu G 2016 *Nano Today* **11** 601-625
[2] Lu Z, Wang H, Kong D, Yan K, Hsu P, Zheng G, Yao H, Liang Z, Sun X, and Cui Y 2014 *Nat. Commun* **5**
[3] Roberts G 2000 *Rev. Process Chem. Eng.* **3** 151-174
[4] Silambarasan M, Padmanathan N, Ramesh P, and Geetha D 2016 *Mater. Res. Express* **3** 095021
[5] Liu S, Hui K, and Hui K N 2016 *Appl. Mater. Interfaces* **8** 3258-67
[6] Wang Y, Shen C, Niu L, Li R, Guo H, Shi Y, Li C, Liu X, and Gong Y 2016 *J. Mater. Chem.* A. **4** 9997-9985
[7] Wanga Y, Yanga D, Zhou T, Pan J, Wie T, and Sun Y 2017 *Nanotechnology* **28** 465402
[8] Dutta A, Dutta S, Mehetor K, Mondal S, Pal I, and Pradhan U 2016 *Chem. Mater.* ***28*** 1872–1878
[9] Karmakar A and Srivastava S K 2017 *Appl. Mater. Interfaces* **9** 22378-22387
[10] Han X, Wahl S, Russo P, and Pinna N 2018 *Nanomaterials* **8** 249
[11] Li D, Feng C, Liu H, Guo Z 2015 *Sci. Rep.* **5** 11236
[12] Gupta S and Tripathi M 2012 *Cent. Eur. J. Chem.* **10(2)** 279-294
[13] Zhang S, Hu Z, Liu K, Liu Y, He F, and Xie Q 2015 *Trans. Nonferrous Met. Soc. China* **25** 4052-4062
[14] Jung S, McCrory C, Ferrer I, Peters J, and Jaramillo T 2016 *J. Mater. Chem.* A **4** 3068-3076
[15] Bikkarolla S and Papakonstantinou P 2015 *J. Power Sources* **281** 243-251
[16] Shinagawa T, Garcia-Esparza A, and Takanabe K 2015 *Sci. Rep.* **5** 13801

IOP Conf. Series: Materials Science and Engineering **856** (2020) 012009 doi:10.1088/1757-899X/856/1/012009

Synthesis and antiparkinsonian activity of nanocomposite of chitosan-tripolyphosphate-*Mucuna pruriens* L extract (CS-TPP-MP)

R E Sardjono[1], A N Fauziyah[1], M D Puspitasari[1], I Musthapa[1], F Khoerunnisa[1], G N Azzahra[2], R Mamat[3], Erdiwansyah[3]

[1]Chemistry Department, Universitas Pendidikan Indonesia, Indonesia
[2] Badan Pusat Statistik Indonesia
[3] Mechanical Engineering Faculty, Universiti Malaysia Pahang, Malaysia

ratnaeko@upi.edu;

Abstract. *Mucuna pruriens* L. (MP) has antiparkinsonian activity because it contains levodopa that acts as a dopamine precursor and plays a role to stimulate dopaminergic receptors in Parkinson's sufferers. The therapeutic efficacy of MP extract can be improved by using a nano drug carrier system, such as chitosan-tripolyphosphate (CS-TPP). This study aims to synthesize, characterize, and evaluate antiparkinsonian activity of chitosan-tripolyphosphate-MP extract (CS-TPP-MP) nanocomposite in mice. MP seed powder was extracted by maceration method using water-ethanol (1:1) by adding citric acid until it reached pH 3. The CS-TPP-MP nanocomposite was synthesized by using ionic gelation method with variations in reactant composition and reaction time. The CS-TPP-MP nanocomposite was characterized by Scanning Electron Microscopy-Energy Dispersive X-ray (SEM-EDX), X-Ray Diffraction (XRD) and Fourier-Transform Infrared (FTIR) Spectroscopy. Catalepsy test was performed to find the antiparkinsonian activity level of CS-TPP-MP nanocomposite at doses of 5, 10, 15, 20, and 25 mg/kg body weight. Based on the results of CS-TPP-MP synthesis, it was found that the reactant composition (CS-TPP:MP) of 1:3 with reaction time of 20 minutes produced the highest yield (14.21%.) SEM-EDX characterization showed that the morphology of CS-TPP-MP nanocomposite was predominantly spherical and the size was approximately 120-170 nm with a composition of C = 55.43%, O = 30.46%, N = 13.46%, P = 0.44%. XRD diffractogram showed that CS-TPP-MP nanocomposite has amorphous structure. FTIR analysis showed the appearance of absorption at wavelength of 1643.35 cm^{-1} which proved the interaction between the primary amine group of chitosan and the carbonyl group of EMP. Catalepsy test demonstrated that CS-TPP-MP nanocomposite at the doses of 5, 10, 15, 20, and 25 mg/kg body weight could reduce catalepsy symptoms in mice significantly, and the best dosage was 20 mg/kg body weight.

1. Introduction

Parkinson's disease is a common degenerative disorder of the central nervous system after Alzheimer's disease [1,2]. Parkinson's disease is due to the decrease of dopamine levels, therefore certain nerve cells in the brain break down and cause abnormal movement and emotion [3,4]. One of the main symptoms of Parkinson's disease is catalepsy. Catalepsy is a disorder of the central nervous system characterized by muscular rigidity and fixity of posture regardless of external stimuli, as well as decreased sensitivity to pain [5-8]. To reduce catalepsy, it is advantageous to use synthetic drug, such as carbidopa, and levodopa. However, the drugs have side effects of prolonged use, namely dyskinesia, digestive disorders, orthostatic hypotension, insomnia, liver abnormalities, visual hallucinations, and depression [9-12]. Therefore, it is necessary to discover herbal medicine with minimum side effect. The minimum side effect of herbal medicine is resulted from extensive activity of its chemical constituents and its product [13-18].

The velvet bean plant (*Mucuna pruriens L.*) has been used by the ancient Indians for the treatment of Parkinson's disease as it possesses approximately 7.56-13.9% levodopa [19,20]. Its extract is more effective and has lower toxicity than synthetic drugs [20].

The implementation of nanotechnology for herbal medicine delivery system is a recent approach. The strategy is more beneficial since herbal plant extract is presented in nanoparticle structure (1-200 nm) made from biodegradable and biocompatible polymers; it is chitosan aiming at increasing solubility, bioavailability, and pharmacological activities [21-25].

Chitosan is a natural polysaccharide which can mostly be identified in the exoskeleton Crustacea, insects, and fungus. The molecular structure of chitosan is a linier polysaccharide composed of polymer glucosamine (β-1,4-2-amino-2-deoxy-D-glucose) [26,27,28]. Chitosan, in medical treatment, can be implemented in the regeneration of epithelial tissue, bone and teeth, wound dressing, antimicrobial, antioxidant, and anti-inflammatory. In addition, chitosan can also be considered to be a nanocarrier for some drug delivery systems. Furthermore, there are also some supportive data on biocompatibility of chitosan related to its function in nanomedicine [26,29]. For the specific reason, it is necessary to synthesize the chitosan-tripolyphosphate and MP extract (CS-TPP-MP) nanocomposite. The result will be beneficial as a drug for Parkinson's disease, specifically to reduce the symptoms of catalepsy in *Mus musculus*

2. Experimental section.

2.1 The Synthesis of CS-TPP-MP Nanocomposite

CS-TPP-MP nanoparticles were synthesized by employing the modified ionic gelation method [30]. In summary, 16 mL of tripolyphosphate solution (TPP) 0.07% (w/v) was added successively into 15 mL of chitosan solution (CS) 0.2% (w/v) then homogenized by magnetic stirrer. MP seed extract (10,000 ppm) was added successively into CS-TPP. Each solution was processed to sonication for 30 minutes, both before and after mixing process. The optimization of synthesizing CS-TPP-MP nanocomposite was performed on variations of CS-TPP and EMP reagent compositions of 1: 3, 1: 2, 1: 1, 2: 1, and 3: 1. It was also performed at some reaction time; 5, 10, 20, 40, and 60 minutes.

CS-TPP-MP solution was into shaker process for 12-15 hours at room temperature then centrifuged at 4000 ppm for 90 minutes. The formed residue was washed by aqua, moved on demineralization process, and dried at room temperature then, finally stored in desiccators. The synthesis results were identified by SEM-EDX (JSM-IT300 InTouchScope) to determine size, product morphology, and chemical composition of the product. To discover the functional groups, the analysis was performed by FTIR spectroscopy (Prestidge 21 Shimadzu). The X-ray diffraction patterns were obtained by using an X-ray diffractometer (Panalytical Xpert) operated at 40 kV and 30 mA with Cu-Kα radiation (k = 1.54060Å).

2.2 The Catalepsy Test

The catalepsy test was performed by a modified bar test method [31,32,33]. It was regarded with placing the animal test in hanging position on its two front legs, after neuroleptic treatment, namely haloperidol.

The animal test was male mice (*Mus musculus*) in age of three months with 23-38 grams as their weight. The mice were kept at standard condition ± 22° C in polypropylene cage for approximately one week to adapt to laboratory condition. CP 551 and mineral water were fed into the mice. The mice were distributed into nine different groups. Each group consisted of three animals. The nine groups were categorized as normal, negative control, positive control, and experimental group (CS-TPP-MP dose at 5, 10, 15, 20, 25 mg/kg of body weight and MP extract 200 mg/kg of body weight).

The intensity of the catalepsy was measured as the length of time for the mice in hanging position on their two front legs holding a wire with 0.5 cm as its diameter and 8 cm as its height without doing any moves. The catalepsy observation was performed after giving haloperidol suspension 5 mg/kg of body weight at interval of 30-60 minutes. Haloperidol was transformed to the mice intramuscularly in 30 minutes after oral treatment of carrier suspension (PGA 1%), levodopa, MP extract or CS-TPP-MP at each predetermined dose.

IOP Conf. Series: Materials Science and Engineering **856** (2020) 012009 doi:10.1088/1757-899X/856/1/012009

The data was processed statistically using the One-Way Analysis of Variance (Anova) followed by Post Hoc Tukey test to evaluate significant differences between the groups. The data was analysed using SPSS 20 software and was considered significant at $P<0,05$.

3. Result and discussion

The used extraction method was maceration since it does not require heating process during the extraction. Therefore, the thermolable compound, as a major compound in MP seed, will never be damaged [34] . A research conducted by Misra and Wagner [35] shows that levodopa dissolves well in water-ethanol (1:1) which is acidified by the addition of citric acid to pH 3. The addition of citric acid aims at determining the pH. The use of ethanol aims at performing membrane cells denaturation, so that the extraction process of active compounds is more effective [36]. The dried extract obtained from the extraction of tangible black solid is 161.3 grams with the percentage of yield was 2.68%.

The Synthesis and Characterization of CS-TPP-MP Nanocomposite

The formation of CS-TPP-MP nanocomposite was performed by the change in color of CS-TPP solution, from colourless to deep black after being mixed by MP extract and put in sonication process. Ultrasonic waves (>20 kHz) can break down aggregates or agglomerates as well as polymer degradation through a process known as cavitation to produce small particles [37,38]. The adjustment of the parameters of reactant composition and reaction time is necessary to determine the optimum conditions based on the percentage of yields.

Variations in reactant composition, in fact, does not result a significant change to the percentage of yield. The reactant composition (CS-TPP: MP) 1:3 results the highest percentage compared to other reagent compositions. It has resulted product of 14.21%. This shows that more MP extract products are trapped in CS-TPP-MP nanocomposite.

The formation of nanocomposite which is an interaction between CS-TPP and MP extract, in fact, depends on the reaction time. The number of yields of CS-TPP-MP nanocomposite increases in the reaction time of 5 to 20 minutes, but at 20 to 60 minutes, the yields decreases. This confirms the previous experiment conducted by Buranachai *et al.* [39] stating that longer reaction can reduce the percentage of synthesis result as the swelling process is possible to release the filler that initially already entangled.

The result of SEM characterization (Fig 1) shows that the morphology of CS-TPP-MP nanocomposite is spherical with 120-170 nm in size. The respective size of CS-TPP-MP nanocomposite is at the criteria for nanoparticles; it is in the range of 100-200 nm [21]. Based on the EDX result in Fig 1, it is known that the CS-TPP-MP nanocomposite contains of elements C, O, N and P with the percentage of 55.43%; 30.46%; 13.46% and 0.44%. This shows the fixed formation of composite between chitosan, tripolyphosphate and MP extract. The existence of chitosan and MP extract is indicated by the presence of elements C, O, and N, while the existence of tripolyphosphate is indicated by the existence of element P. However, the data cannot be sure to confirm the composition of CS-TPP-MP nanocomposite.

IOP Conf. Series: Materials Science and Engineering **856** (2020) 012009 doi:10.1088/1757-899X/856/1/012009

ZAF Method Standardless Quantitative Analysis
Fitting Coefficient : 0.4794

Element	(keV)	Mass%	Sigma	Atom%
C K	0.277	55.43	1.19	61.48
N K	0.392	13.46	2.56	12.80
O K	0.525	30.67	2.33	25.53
P K	2.013	0.44	0.26	0.19
Total		100.00		100.00

Figure 1. SEM visualization of CS-TPP-MP nanocomposite

The functional groups discovered in MP extract and CS-TPP-MP nanocomposite can be characterized by FTIR spectroscopy. Based on FTIR CS-TPP-MP, the typical *schiff base* exists at wave number 1632, 53 cm^{-1} that shows the interaction between the primary amine group of chitosan and the carbonyl group of MP extract. FTIR CS-TPP and CS-TPP-MP spectra show the absorption group of P=O at wave number 1150 cm^{-1}. The difference in intensity and shift in wave numbers are due to the decrease of hydrogen bonds. It shows that there is an interaction between MP extract and chitosan-tripolyphosphate [40].

The X-Ray Diffraction (XRD) was performed to identify the crystal structure of CS-TPP-MP nanocomposite. The diffractogram shows that the CS-TPP-MP nanocomposite possesses low-crystallinity and is amorphous (random and irregular pattern) so that the identified peak point of 2θ is different from literature, namely the absorption for CS-TPP is in the range of 20.20°, CS-TPP-LD is in the range of 8.06°, 16.04° and 24.52°, and the diffraction peak 2θ for the typical levodopa will exist at 16.04° [41,42]. In addition, the peak absorption of 2θ for CS-TPP-MP nanocomposite is at 54.10° and 69.38°. Amorphous crystal structure is resulted from the high composition of MP extract in CS-TPP-MP nanocomposite.

Catalepsy Test

The method was the bar method regarded with the observation of mice response when they are in hanging position on their two front legs holding a wire with 0.5 cm as its diameter and suspended 8 cm from the surface [31]. Mice are considered to experience catalepsy when they are in hanging position more than 15 minutes.

The negative control group (haloperidol treatment at 5 mg/kg of body weight) shows that the hanging time is more than 20 seconds. Therefore, the mice experience catalepsy. On the other hand, the positive control group (levodopa treatment dose at 10 mg/kg of body weight), MP extract 200 mg/kg of body weight, and CS-TPP-MP nanocomposite at 5, 10, 15, 20, 25 mg/kg of body weight show the hanging time is less than 20 seconds. It means that the mice do not experience catalepsy. Furthermore, the respective group has a longer hanging time than normal control group (1.55 ± 0.33 seconds). It shows that the positive control group (L-dopa treatment dose at 10 mg/kg of body weight), MP extract 200 mg/kg of body weight, and CS-TPP-MP nanocomposite 5, 10, 15, 20, 25 mg/kg of body weight are able to reduce the symptoms of catalepsy but have not yet reached normal condition. In addition, the hanging time for the group of mice with CS-TPP-MP nanocomposite doses of 5, 10, 15, 20, 25 mg/kg of body weight is smaller than the group of mice with MP extract. A prior research conducted by Sardjono *et al.* [43] shows the identical result. The antiparkinsonism activity of nanoparticles extract is more effective than MP extract. The hanging time of mice treated by CS-TPP-MP nanocomposite with dose at 20 and 25 mg/kg of body weight does not have significant change. It is possible as the optimum dose in mice is at 20 mg/kg of body weight.

The *one-way* ANOVA statistical analysis continued by *Post Hoc* Tukey test were performed to determine the significance of CS-TPP-MP nanocomposite in reducing catalepsy symptoms. The used significance limit is 0.05 with the confidence level of 95%. If the P value is less than 0.05, then the data show a significant difference. Based on the results of statistical tests, it shows that CS-TPP-MP doses of 5, 10, 15, 20, 25 mg/kg of body weight have a significance value of less than 0.05 to negative control groups. It shows that the hanging time of mice groups with CS-TPP-MP doses of 5, 10, 15 20, 25 mg/kg of body weight has a significant difference to negative controls. This proves that CS-TPP-MP nanocomposite doses of 5, 10, 15, 20, 25 mg/kg of body weight can reduce the symptoms of catalepsy.

The result of comparison of the hanging time of mice groups with CS-TPP-MP nanocomposite doses of 5, 10, 15 20, 25 mg/kg of body weight to normal control has a significance value less than 0.05. It indicates that the hanging time of mice groups with CS-TPP-MP nanocomposite doses of 5, 10, 15 20, 25 mg/kg of body weight have a significant difference to normal controls. This proves that CS-TPP-MP nanocomposite doses of 5, 10, 15, 20, 25 mg/kg of body weight can reduce the symptoms of catalepsy but have not yet reached normal condition.

Conclusion

Based on the results of CS-TPP-MP nanocomposite synthesis, it can be concluded that the composition of CS-TPP: MP reagents is 1: 3 and the reaction time is for 20 minutes. It results the highest percentage of yields, 14.21%. CS-TPP-MP nanocomposite has spherical with 120-170 nm. The identified compositions are C=55, 43%, O=30, 46%, N=13,46%, and P=0,44%. The catalepsy test shows that CS-TPP-MP nanocomposite is able to significantly reduce the symptoms of catalepsy with the best dose of 20 mg/kg of body weight.

4. References

[1] Dickson D W, 2017, *Parkinsonism Related D*, 46,1, S303–S303
[2] Homayoun H, 2018, *Ann Intern Med*, 169(5), ITC33-ITC4
[3] Xu L, Pu J, 2016, *Parkinson Dis*, 4, 1–10
[4] Nomoto M, Nagai M, Nishikawa N, Ando R, Kagamiishi Y, Yano K, Saito S, Takeda A,2018, *eNeurologicalSci* ,13,8-13
[5] Sonegoa A B, Gomesa F V, Belb E A D, Guimaraesa F G, 2016, *Behav Brain Res*, 309,22-28
[6] Ionov I D, Pushinskaya I I, Gorev N P, Frenkel D D, 2018, *Neurosci Lett*, 684,72-77
[7] Colombo A C, de Oliveira A R, Reimer A E, Brandão M L, 2013, *Behav Brain Res,* 257,201-207
[8] Banasikowski T T, Beninger R J, 2012, *Eur Neuropsychopharmacol,* 9, 22,761-768
[9] Brooks D J, 2008, *Neuropsyc Dis Treat*, 4,1,39–4
[10] Ray S, Agarwal P, 2019 *Clin Geriatri Med*, Article in Press,
[11] Mancini M, Rocchi L, Horak F B, Chiari L, 2008, *Clin Biomechanics*, 23,4,450-458
[12] Bermejo D D, Campo M V, o López Y M, 2019, *Enfermeria Clin*, article in Press
[13] Mills, S. ,1996, *Pengobatan Alternatif (Alternative in Healing)*. Jakarta: Dian Rakyat
[14] Bao X-X, Ma H-H, Ding H, Li W-W, Zhu M, 2018, *J Integr Med*,16,290-296
[15] Kim D H, Choi J J, Park B J, 2019, *Integr Med Res,*8,2019,202-208
[16] Song J-X, Sze S C-W, Ng T-B, Lee C K-F, Leung G P H, Shaw P-C, Tong Y, Zhang Y B, 2012, *J Ethnopharmacol,*139,698–711
[17] Li X-z, Zhang S-n, Liu S-m, Lu F, 2013, Fitoterapia, 84,273–285
[18] Chen K-Y, Wu M-Y, Yang P-S, Chiang J-H, Hsu C-Y, Chen C-Y, Yen H-R, 2018, *J Ethnopharmacol*,226,168-175
[19] Sardjono R E, Musthapa I, Sholihin H, Ramdhani R P, 2012, *GJRMI*, 4,101–108
[20] Maldonado R G, 2018, *Understanding Pathophysiology and Developing Therapeutic Strategies*, 95–116

[21] Chen M, Zhou X, Chen R, Wang J,Ye R D, Wang Y, Wu C, Mahato R I, 2019, *Mater Today*, 25, 66-87
[22] Harwansh R K, Deshmukh R, Rahman M, 2019, *J Drug Deliv Sci Tec*, 51, 224-233
[23] Saraf A S, 2016, *J Controlled Release*, 241,110–124
[24] Nagalingam A, 2017, *Japanese Kampo Medicines for the Treatment of Common Diseases: Focus on Inflammation :Chapter 15 - Drug Delivery Aspects of Herbal Medicines*, 143-164
[25] Qamar S A, Asgher M, Khalid N, Sadaf M, 2019, *Biocatal Agri Biotech,* 22,101388 ,1-10
[26] Maa X, Fenga H, Lianga C, Liua X, Zenga F, Wang Y, 2017, J Mat Sci Tech,33,1067–1074
[27] Li J, Hwang I-C, Chen X, Park H J, 2016, *Food Hydrocolloids*, 60, 138e147
[28] Yu S, Xu X, Feng J, Li M, Kaili H, 2019, *Int J Pharm*, 560, 282-293
[29] Asgari-Targhi G, Iranbakhsh A, Ardebili Z O, Asgari-Zahra G, Ardebili O, 2018, *Plant Physiol Bioch,*127 393–402
[30] Anandhakumar S, Krishnamoorthy G, Ramkumar K M, Raichur A M, 2017, *Mat Sci Eng C*, 70 378–385
[31] Sanberg P R,Emerich D F, El-Etri M M, Shipley M T, Zanol M D, Cahill D W,Norman A B, 1993, *Pharmacol Biochem Be*, 46,2,303-307
[32] Sardjono R E, Khoerunnisa F, Musthopa I, Qowiyah A, Khairunisa D D, Erfianty D, Rachmawati R, 2018, *JESTEC*,13,12,4258–4270
[33] Ionov I D, Pushinskaya I I, Gorev N P, Frenkel D D, 2018, *Neurosci Lett*, 684,72–77
[34] Zhang Q W, Lin L G, Ye W C, 2018, *Chin Med,*13, 20,1–26
[35] Misra L, Wagner H, 2007, *Ind J Biochem Biophys*, 44,1,56–60.
[36] Hermawati Y, Rofieq A, Wahyono P, 2015, Proceeding *Seminar Nasional Pendidikan Biologi*, 301–308
[37] Taurozzi J S, Hackley V A, Wiesner M R, 2010, *Nanotoxicology*, 5,4, 711–729
[38] Antoniou J, Liu F, Majeed H, Qi J, Yokoyama W, Zhong F, 2015, *Colloids Surf,* 465, 137–146
[39] Buranachai T, Praphairaksit N, Muangsin N, 2010, *AAPS PharmSciTech*, 11,3,1128-1137
[40] Ryu S O O R, Noda I, Jung Y M E E, 2010, *App Spectrosc*, 64, 9, 1017–1021
[41] Anand M, Sathyapriya P, Maruthupandy M, Beevi H, 2018, *Front Lab Med*, 2, 2, 72–78
[42] Shehabeldine A, Hasanin M, 2019, *Environ Nanotech, Monitor Man,*12,100252,1-8
[43] Sardjono R E, Khoerunnisa F, Musthopa I, Akasum N S M M, Rachmawati R, 2018, J Phys: Conference Series, 1013, 012195,1-8

IOP Conf. Series: Materials Science and Engineering **856** (2020) 012010 doi:10.1088/1757-899X/856/1/012010

Modeling and simulation of a gauge shaped beam coupled with macro fiber composite for energy harvesting application

D S Ibrahim[1,2], B B* Sun[1], U Sharif[1] and O A Oluseyi[1]

[1] School of Mechanical Engineering, Southeast University, Nanjing, People's Republic of China
[2] Department of Mechatronic Engineering, Faculty of Engineering, Bayero University, Kano, P.M.B 3011, Nigeria

bbsun@seu.edu.cn

Abstract. This paper presents a novel unimorph piezoelectric cantilever with non-conventional geometry named as gauge shaped beam to investigate stress, Eigen frequencies and electric potential for improved vibration energy scavenging. The harvester is made of macro fiber composite (MFC) as the active piezoelectric layer bonded to a brass substrate. Complete transversely isotropic elastic and piezoelectric properties are assigned to the composite structure in a three-dimensional finite element model using COMSOL Multiphysics. Dynamic behaviour of the proposed structure was simulated and the results obtained were compared with traditional rectangular geometry. The results proved that the gauge shaped beam has a higher stress distribution around the gauge length, enhanced electric potential and lower resonance frequencies than the conventional rectangular structure which makes it better for the design of piezoelectric based vibration energy harvesters.

1. Introduction

In a numerous effort to achieve 100% pollution free environment by most countries of the world, ambient energy scavenging from unconventional sources such as Hydro, solar, wind, and geothermal have attracted keen research interest from various field of studies[1-3]. The aforementioned technologies have been successful in providing green energy in macro scale. But, when dealing with power generation in micro scale, the readily available technology is using chemical batteries. However, batteries have stringent lifetime and there is difficulty associated with their replacement when placed at remote or inaccessible locations. Moreover, they are not biodegradable which make disposal of the depleted chemical cells infeasible. Thus, scavenging mechanical energy in form of vibration using piezoelectric (PZT) transducers has become one of the potential technologies to replace the battery cells. Piezoelectric material bonded to a rectangular cantilever beam as a unimorph or bimorph has been the generic configuration used for designing harvesters [4]. The rectangular cantilever beam serving as the substrate structure has the following advantages: For a given input excitation, it produces average mechanical strain and also it can easily be manufactured through micro-fabrication process. In spite of this, the output from the layered active material is usually below the threshold needed to power the intended systems [5]. This could be attributed to non-optimal utilization of the strain experienced on the shim. Therefore, to enhance the conversion efficiency and thus the energy scavenging abilities of piezoelectric energy harvester (PEH) systems, development of innovative configurations have been considered as one of the premium solutions for optimum vibration based energy harvesting [6-8]. In this paper, a gauge shaped cantilever beam is designed and simulated with COMSOL Multiphysics 5.3 for the conversion of vibration energy into electrical energy via piezoelectric transducer to achieve a larger strain and higher voltage at lower frequency.

2. Piezoelectricity Theory and Governing Equations

Piezoelectric materials produce charges when subjected to mechanical deformation by applied force or vibration. In addition, it undergoes strain when voltage is applied to the material. Therefore, piezoelectric material exhibit a two way coupling which can be described by constitutive equations in a compacted matrix form as [9];

$$D = \varepsilon^T \underline{E} + d'\underline{T} \quad (1)$$

$$S = s^E \underline{T} + d'\underline{E} \quad (2)$$

Where D represents the electric displacement, ε^T is the dielectric permittivity tensor, $\underline{E}$ is the applied electric field vector, $\underline{T}$ is the stress vector, S is the strain vector, s^E is the elastic compliance matrix and d' is the piezoelectric coefficient or Electro-mechanical coupling factor.

One of the fundamental factors that influences the generation of charges in a piezoelectric cantilevered beam is the amount of stress produced when loaded. Hence, a cantilevered PEH subjected to base excitation in 1st mode of vibration, produces voltage as a function of the stress experienced in the material as[10] ;

$$V_{31} = \sigma_{xx} g_{31f} t_{xx} \quad (3)$$

Where σ_{xx} is the applied stress, g_{31f} is the effective piezoelectric constant and t_{xx} is the thickness of PZT layer. Generally, ambient vibration sources exhibit lower frequencies below 200 Hz [11, 12]. In order to harness the mechanical vibration properly, resonant frequency of the PEH should be less than the range. Thus, optimum energy can be harnessed efficiently when harvester is driven at resonant frequency .Due to flexibility, lower resonant frequency and high stress generation cantilever beam structure is more preferred [13].

3. Design of Gauge Shaped and Rectangular beam Harvesters

The proposed gauged shaped harvester was designed using SolidWorks. It is made of three components; *a substrate*, *macro fiber composite* (MFC) as the piezoelectric layer and *adhesive* for bonding the smart material. Figure 1(a) and (b) show the schematics of the harvesters. The dimensions of the substrates are shown in Figure 2(a) and (b) while the dimension (length x width x thickness in mm) of MFC 2807-P2 (28 x 7 x 0.3) from smart material corporation [14] is considered in the design. The adhesive has the same dimension with the MFC patch except its thickness, which is considered to be 0.1mm. It is important to note that the design was aimed at maintaining equal mass for fair comparison.

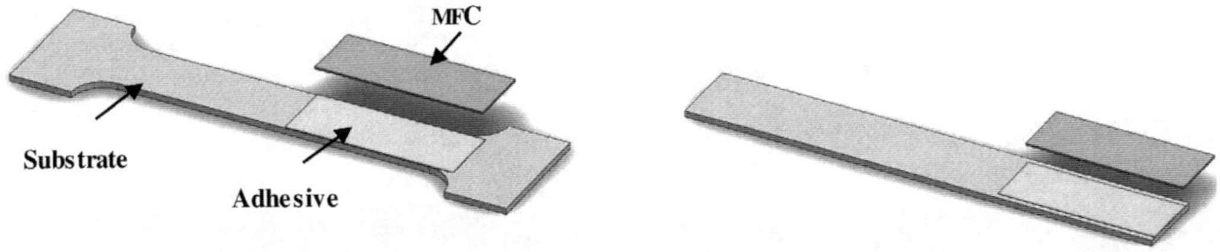

(a) (b)

Figure 1. (a) Proposed gauge shaped beam harvester and (b) Conventional rectangular beam harvester.

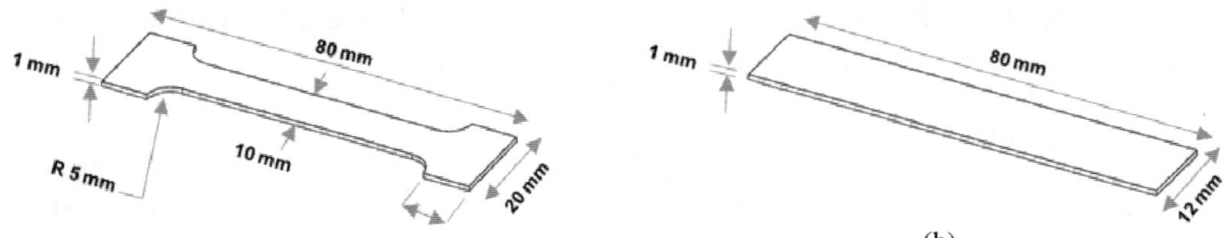

(a) (b)

Figure 2. (a) Substrate dimensions of the proposed gauge shaped and (b) Conventional rectangular beam.

4. Modeling in COMSOL Multiphysics

The designed 3-D models of the substrates were exported to COMSOL to carry out static analyses. Brass was chosen as the substrate materials and the boundary conditions for the beams such as fix end, free end and applied static force (20N) were set in solid mechanics physics node of the FE model builder. Subsequently, the adhesive and the MFC patch were coupled with the substrate using assembly operation in SolidWorks and then exported again into the FE software to analyse its mechanical and electrical properties. Table 1 shows the physical properties of brass and adhesive used for the simulation. To fully define the MFC patch properties in COMSOL Multiphysics, we need the orthotropic elastic compliance (S^E), dialectic piezoelectric constant (d') and dielectric permittivity (ε^T) matrices. Therefore, S^E and d' matrices were populated using manufacturer's data sheet values [14] whereas; the ε^T was obtained using the mixing rule for the $\varepsilon_{33}{}^T$ by Deraemaker et.al. [15] and a mixing rule for dielectric composites $\varepsilon_{22}{}^T$ from Jaffe [16]. After that, the top and bottom surfaces of the piezoelectric layers were defined as ground and terminal in boundary conditions of electrostatics physics. The rest of the faces were labelled as zero charge. In the next step, a tetrahedral mesh with extremely fine element size was used for the discretization of the beams. Finally, Eigen frequency and electric potential analysis were performed to determine eigenvalues and voltage output potential of the harvesters.

Table 1. Properties of materials used for simulation

Parameters	Brass	Adhesive (Epoxy DP-460)
Young's Modulus (N/m^2)	100x10^9	2.7x10^9
Poison's ratio	0.34	0.4
Density (Kg/m^3)	8740	1100

5. Simulation Results and Discussion

In this section, the two configurations have been studied to obtain their mechanical and electrical characteristics. Figure 3 (a) and (b) show the static analysis results. From the colour contour of von Mises stress for the gauge shaped beam, a maximum stress of 9.29x10^8 N/m^2 was observed around the curvature region close to the face where it was fixed.

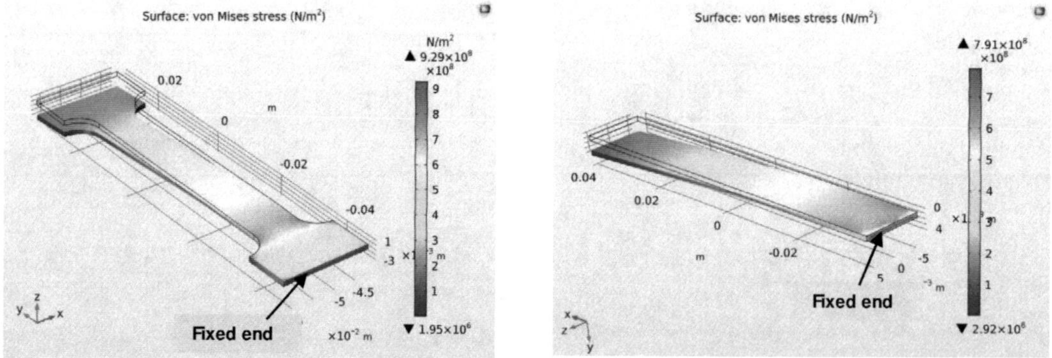

(a) (b)

Figure 3. Von Mises stress analysis of (a) Gauge shaped beam and (b) Rectangular cantilever beam.

IOP Conf. Series: Materials Science and Engineering **856** (2020) 012010 doi:10.1088/1757-899X/856/1/012010

On the other hand, the conventional beam has a maximum stress of 7.91x10^8 N/m^2 close to its clamped end. Hence, it can be deduced that the energy conversion from mechanical to electrical will be higher when a piezoelectric cantilevered beam experienced larger stress at its resonance frequency. Therefore, it is obvious that the proposed beam harvester has greater tendency of converting its stress into voltage output based on relationship presented in equation (3). It is worth noting that the lower values of stresses (1.95x10^6 N/m^2 and 2.92x10^6 N/m^2) exhibited around the free ends have less significance on the energy conversion, which is why the design of partially covered piezoelectric beam harvesters are mostly considered in the literature. Furthermore, an Eigen frequency analysis was conducted to determine resonance frequencies as well as tip displacements of the two configurations. Figure 4 (a) and (b) show the first three vibration modes of the gauged shaped and rectangular cantilever beam harvesters respectively.

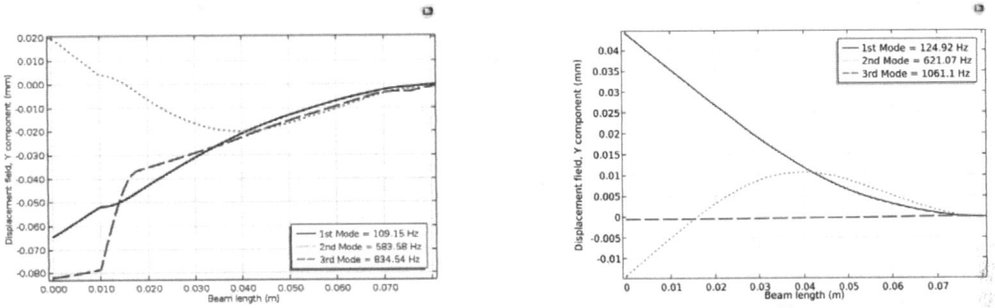

(a) (b)

Figure 4. Displacement of (a) Gauge shaped and (b) Conventional beam harvester at different Eigen frequencies.

The results show that the free end of the gauged shaped harvester was displaced and reached a maximum value of 0.065 mm at a fundamental resonance frequency of 109.15 Hz. Also, maximum tip displacement values of 0.02 mm and 0.08 mm were recorded at second and third resonance frequencies of 583.58 Hz and 834.54 Hz respectively. For the conventional harvester, maximum tip displacement of 0.045 mm, 0.001 mm and 0.015 mm were obtained at 1st, 2nd and 3rd resonance frequencies of 124.92 Hz, 621.07 Hz and 1061.1 Hz respectively. It is important to note that there is decrease in natural frequencies of the proposed harvester over the rectangular beam harvester in all the three modes of vibration considered. This advantage is highly desirable in designing vibration based energy harvester due to the fact that most vibration sources have lower excitation frequencies. The lower resonance frequencies of the gauge shape harvester could be attributed to the designed flange at its free end serving as an integrated tip mass. Figure 5 (a) and (b) show the electric potential of the two configurations at their fundamental resonance frequencies (109.15 Hz and 124.94 Hz). A voltage output ranging between 13.3V to -13.2V was achieved from the proposed harvester while 10V to -10V was produced by the conventional harvester. The enhanced output of the gauge shaped harvester could ascribe to the two flanges at both ends which provide an appreciable strain along its length.

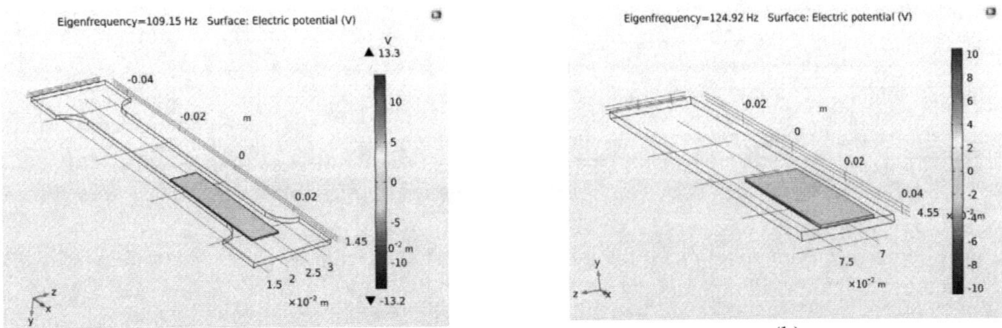

(a) (b)

Figure 5. Electric potential at fundamental resonance frequency of (a) Gauge shaped harvester and (b) Conventional harvester.

IOP Conf. Series: Materials Science and Engineering **856** (2020) 012010 doi:10.1088/1757-899X/856/1/012010

6. Conclusion

In this study, an energy harvester with non-traditional geometry was designed and modelled in COMSOL Multiphysics. The simulation of the proposed geometry (stress, resonant frequency, displacement and electric potential) were analysed and compared with rectangular shaped cantilever beam. The results showed that the proposed shape had a higher stress distribution, lower resonant frequencies and higher electric potential than the conventional beam harvester. Hence, the design of the proposed gauge shaped beam provides a new insight for development of an improved piezoelectric based vibration energy harvester.

References

[1] Yildiz F 2009 Potential Ambient Energy Harvesting Sources and Techniques *The journal of Technology studies* vol. 35, no. 1.

[2] Vatansever D, Hadimani R L, Shah T, and Siores E 20011 An investigation of energy harvesting from renewable sources with PVDF and PZT *Smart Materials and Structures,* vol. 20, no. 5, pp. 055019.

[3] Anzalchi A and Sarwat A 2017 Overview of technical specifications for grid-connected photovoltaic systems *Energy Conversion and Management,* vol. 152, pp. 312-327.

[4] Ibrahim D S, Muthalif A G A, Nordin N H D and Saleh T 2016 Comparative study of conventional and magnetically coupled piezoelectric energy harvester to optimize output voltage and bandwidth *Microsystem Technologies*, pp. 1-12.

[5] Roundy S and Wright P K 2004 A piezoelectric vibration based generator for wireless electronics *Smart Materials and Structures* vol. 13 no. 5, pp. 1131-1142.

[6] Yang Z, Wang Y Q, Zuo L, and Zu J 2017 Introducing arc-shaped piezoelectric elements into energy harvesters *Energy Conversion and Management* vol. 148 no. Supplement C, pp. 260-266.

[7] Siddique A R M, Mahmud S and Van Heyst B 2016 Energy conversion by 'T-shaped' cantilever type electromagnetic vibration based micro power generator from low frequency vibration sources *Energy Conversion and Management* Vol. 133, pp 399-410.

[8] Karami M A and Inman D J 2012 Parametric Study of Zigzag Microstructure for Vibrational Energy Harvesting *Journal of Microelectromechanical Systems,* vol. 21, no. 1, pp. 145-160.

[9] Leo D J 2007 *Engineering Analysis of Smart Material Systems* (New Jersy: John & Sons, Inc).

[10] Priya S, Song H C, Zhou Y, Varghese R, Chopra A, Kim S G, Kanno I, Wu L, Dong S Ha, Ryu J, and Ronald G P 2017 A Review on Piezoelectric Energy Harvesting: Materials, Methods, and Circuits *Energy Harvesting and Systems* p. 3.

[11] Roundy S, Wright P K, and Rabaey J A 2003 study of low level vibrations as a power source for wireless sensor nodes *Computer Communications,* vol. 26, no. 11, pp. 1131-1144.

[12] Liu H, Quan C, Tay C J, Kobayashi J T, and Lee C 2011 A MEMS-based piezoelectric cantilever patterned with PZT thin film array for harvesting energy from low frequency vibrations *Physics Procedia,* vol. 19, pp. 129-133.

[13] Sodano H A, Park G and Inman D 2004 Estimation of electric charge output for piezoelectric energy harvesting *Strain,* vol. 40, no. 2, pp. 49-58.

[14] Smart Material Corp. *Macro fibre Composite* Datasheet Retrieved on 19/5/2019 available at http://www.smart-material.com/mFc-productmain.html,

[15] Deraemaeker A, Nasser H, Benjeddou A, and Preumont A 2009 Mixing Rules for the Piezoelectric Properties of Macro Fiber Composites *J Intel Material Syst Struct,* vol. 20, pp. 1475-1482.

[16] Jaffe H and Berlincourt D A 1965 Piezoelectric transducer materials *Proc. of the IEEE,* vol. 53, no. 10, pp. 1372-1386.

AUTHOR INDEX